PYROGENS AND FEVER

A Ciba Foundation Symposium

Edited by
G. E. W. WOLSTENHOLME
and
JOAN BIRCH

CHURCHILL LIVINGSTONE
Edinburgh and London
1971

First published 1971

With 61 illustrations

I.S.B.N. 0 7000 1504 3

Printed in Great Britain

Contents

Membership

Symposium on Pyrogens and Fever held 8th–10th July, 1970

E. Atkins	Department of Internal Medicine, Yale University School of Medicine, 333 Cedar Street, New Haven, Connecticut 06510, U.S.A.
D. R. Bangham	Division of Biological Standards, National Institute for Medical Research, Mill Hill, London, N.W.7
Phyllis T. Bodel	Department of Internal Medicine, Yale University School of Medicine, 333 Cedar Street, New Haven, Connecticut 06510, U.S.A.
P. K. Bondy	Department of Internal Medicine, Yale University School of Medicine, 333 Cedar Street, New Haven, Connecticut 06510, U.S.A.
K. E. Cooper	Division of Medical Physiology, Faculty of Medicine, The University of Calgary, Calgary 44, Alberta, Canada
W. I. Cranston	Department of Medicine, St. Thomas's Hospital Medical School, London, S.E.1
W. S. Feldberg	National Institute for Medical Research, Mill Hill, London, N.W.7
M. J. Grundman	The Radcliffe Infirmary, Oxford
M. Landy	National Institute of Allergy and Infectious Diseases, National Institute of Health, Bethesda, Maryland 20014, U.S.A.
P. Lechat	Faculté de Médecine, Institut de Pharmacologie, 21 Rue de l'école-de-Médecine, Paris 6e, France
P. A. Murphy*	Department of Microbiology, School of Medicine, The Johns Hopkins University, 725 North Wolfe Street, Baltimore, Maryland 21205, U.S.A.
R. D. Myers	Laboratory of Neuropsychology, Department of Psychology, Purdue University, Lafayette, Indiana 47907, U.S.A.
C. H. R. Palmer	School of Pharmacy, City of Leicester Polytechnic, P.O. Box 143, Leicester LE1 9BH
Sir George Pickering	The Master's Lodgings, Pembroke College, Oxford
M. D. Rawlins	Department of Medicine, St. Thomas's Hospital Medical School, London, S.E.1
L. Saunders	The School of Pharmacy, University of London, Brunswick Square, London, W.C.1

*Unable to attend.

P. A. Saxena	National Institute for Medical Research, Mill Hill, London, N.W.7; and Department of Pharmacology, Aligarh Muslim University, Aligarh, India
K. L. Smith	Quality Control, Bio-Assay, Boots Pure Drug Co. Ltd., Nottingham, NG3 3AA
E. S. Snell	Medical Department, Glaxo Laboratories Ltd., Greenford, Middlesex
P. J. Teddy	Osler House, 43, Woodstock Road, Oxford
T. D. Whittet	Department of Health and Social Security, Queen Anne's Mansions, Queen Anne's Gate, London, S.W.1
Elizabeth Work	Department of Biochemistry, Imperial College of Science and Technology, London, S.W.7

Preface

For some 15 years Sir George Pickering, Dr T. D. Whittet and Dr K. E. Cooper had been urging us to hold a Ciba Foundation symposium to review the problems of pyrogens and fever. However, it always seemed that more progress in research was desirable to make such a review profitable. But by 1969 there was evidence, in particular from Dr D. R. Bangham, of an urgent need for new, standardized and generally acceptable methods for the assay of pyrogens. This, together with the persisting pressure from the original quarters, led to this symposium being held in July 1970.

The meeting revealed the need for still further research, and it is encouraging that the discussions gave rise to new lines of work within a matter of weeks. It is hoped that more investigations in this vital area, which affects both clinical treatment and the production of therapeutic agents, will be stimulated by thoughtful study of these papers and discussions by readers in various parts of the world.

The editors are most grateful to all who contributed to the meeting and to the preparation of this volume, particularly those already named above.

The Ciba Foundation

The Ciba Foundation was opened in 1949 to promote international cooperation in medical and chemical research. It owes its existence to the generosity of CIBA Ltd (now CIBA-GEIGY Ltd), Basle, who, recognizing the obstacles to scientific communication created by war, man's natural secretiveness, disciplinary divisions, academic prejudices, distance, and differences of language, decided to set up a philanthropic institution whose aim would be to overcome such barriers. London was chosen as its site for reasons dictated by the special advantages of English charitable trust law (ensuring the independence of its actions), as well as those of language and geography.

The Foundation's house at 41 Portland Place, London, has become well known to workers in many fields of science. Every year the Foundation organizes six to ten three-day symposia and three to four shorter study groups, all of which are published in book form. Many other scientific meetings are held, organized either by the Foundation or by other groups in need of a meeting place. Accommodation is also provided for scientists visiting London, whether or not they are attending a meeting in the house.

The Foundation's many activities are controlled by a small group of distinguished trustees. Within the general framework of biological science, interpreted in its broadest sense, these activities are well summed up by the motto of the Ciba Foundation: *Consocient Gentes*—let the peoples come together.

NOTE ON TERMINOLOGY

E. Atkins

Throughout the discussions, the terms "bacterial pyrogen" and "endotoxin" are often used interchangeably. This usage, though honoured by tradition, is unfortunately misleading and stems from the discovery that the so-called "injection fevers" of the 19th century were due to the inadvertent contamination of biological materials with pyrogens of bacterial origin present in the air and water. These agents, which are now known to be derived from the cell walls of gram-negative bacteria, have been given the more specific name "endotoxins" and have been identified biochemically as lipopolysaccharides of high molecular weight. Bacteria of other classes, e.g. gram-positive bacteria and mycobacteria, do not appear to contain substances with the same biochemical and physiological properties as endotoxins, although these bacteria may contain or produce a number of agents that will cause fever when given intravenously and therefore should properly be included in the older term "bacterial pyrogens". In a number of instances, these substances are proteins and are presumably pyrogenic by virtue of antigen-antibody reactions occurring in naturally or specifically sensitized hosts.

It is apparent, therefore, that "bacterial pyrogen", as opposed to "gram-negative bacterial endotoxin", is a general descriptive term for many different substances. Since other microbial agents may also be pyrogenic (e.g. certain viruses and fungi) it would seem preferable to retain the term "bacterial pyrogen" only in this generic sense and not as a substitute for endotoxin.

These microbial substances as a group may be referred to as examples of "exogenous pyrogens" to contrast them with pyrogens derived from the tissues of the animal host which are now known collectively as "endogenous pyrogens" or "leucocyte pyrogens".

"Leucocyte pyrogen" is a term usually reserved for the pyrogenic agent isolated from either blood or exudate leucocytes. The major cell type in these instances is the granulocyte although, as will be apparent from the papers and discussions in this symposium, other cell types, e.g. monocytes, macrophages and Kupffer cells, also produce a pyrogen, the biochemical nature of which has not yet been worked out in any detail.

CHAIRMAN'S OPENING REMARKS

Sir George Pickering

It is nearly forty years since I did my first piece of entirely original research. Lewis and I worked on peripheral vascular disease and one of the problems was to identify and measure obstruction of the main arteries. Lewis thought that if we heated the body the extremities would get hot by vasodilatation which might be measured through skin temperature. We made a chamber to enclose the trunk. When the chamber was heated the skin temperature of the extremities rose, but it did not rise as high or as quickly when there was arterial obstruction (Lewis, Pickering and Rothschild, 1931). The method is useful and still persists.

What interested me more was the mechanism by which the effect was produced. Was it a reflex from the skin, as was currently thought, or was it the effect of warm blood on a central mechanism? I showed that when one limb with its circulation arrested is plunged into cold water there is almost immediate vasoconstriction in the other limb. This wears off. When the circulation is released the vasoconstriction returns after a latent period and persists as long as immersion is continued. When the arm with arrested circulation is plunged into warm water there is no change in blood flow in the opposite limb. When the circulation is released there is no change for several minutes and then vasodilatation begins, increases, and continues as long as the limb is immersed. It seemed quite clear that the application of cold to the skin evoked a vasoconstrictor reflex, but the application of warmth did not evoke a vasodilatation reflex, though both heat and cold had an effect on inducing vasodilatation and vasoconstriction respectively (Pickering, 1932).

This was further investigated by Snell in my department. He showed that infusing warm or cold saline into the antecubital vein would produce vasodilatation or vasoconstriction in the opposite hand. The size of the response was directly related to the amount of the heat transfer. When the responses to raising and reducing the arterial blood temperature were plotted the points lay on a single line. The central receptor was very sensitive, responding to changes of less than 0·1°C (Snell, 1954).

To locate the central receptor, Downey, Mottram and I (1964) cooled single arteries and veins in the conscious rabbit and measured the increase in oxygen consumption. By far the most responsive area was that served by the internal carotid artery. It seems likely, therefore, that the receptor lies in that territory, in agreement with the observations on direct heating and cooling made by Ström (1950).

When man exercises vigorously the body temperature rises, sometimes by as much as 4°C. This is not a fever. The rise is accompanied by vasodilatation and sweating which, when exercise is discontinued, reduce the temperature to more or less its previous level. Fever is not a simple alteration in heat gain or heat loss. It represents a disturbance of the central mechanism regulating body temperature so that the temperature is raised. Von Liebermeister (1875) thought that fever was due to a change in setting of the central mechanism. Evidence of this was produced by Stern (1892) by rather crude methods. Much better evidence was found by Cooper, Cranston and Snell (1964), who showed that a given heat transfer produced much the same vasodilatation in a given patient when his temperature was normal or raised to a set level by injection of a pyrogen.

Gerbrandy, Cranston and Snell (1954) showed that when bacterial pyrogen was injected intravenously, cutaneous vasoconstriction and increase in mouth temperature did not begin until about fifty minutes after the injection. If, however, the same dose of pyrogen was incubated with 200 ml of blood for two hours and then injected, the latent period was reduced to about twenty minutes. This suggested that bacterial pyrogen first exerts its effect in the blood. Analysis showed that this was due to white cells, not red cells, platelets or plasma. Earlier Bennett and Beeson (1953) had shown that the leucocytes are the only tissue in the body from which it is easy to recover a pyrogenic agent. This agent, endogenous or leucocyte pyrogen, was quite different from bacterial pyrogen in that it was destroyed by heat. Thus it seemed likely that fever was produced by bacterial pyrogen causing leucocytes to produce leucocyte pyrogen, and leucocyte pyrogen then acted on some other structure.

Just before I left my department, Cooper and his colleagues showed that minute doses of leucocyte pyrogen injected into the pre-optic region of the anterior hypothalamus, near the midline, produced fever within a few minutes. When injected elsewhere much larger doses were needed and the latent period was longer. Bacterial pyrogen also acts in larger doses and after a longer latent period. It seems, therefore, that the structure on which leucocyte pyrogen acts is in the pre-optic region of the anterior hypothalamus (Cooper, Cranston and Honour, 1966).

Briefly, our current concept of the mechanism of fever is that infection releases leucocyte pyrogen from leucocytes, and perhaps from other cells, and the leucocyte pyrogen then acts on cells in the anterior hypothalamus which regulate body temperature so that their "set-point" is raised. This is a hypothesis which can be refuted by further observation and experiment. And arising from it, I would like to ask three questions:

(1) Is endogenous pyrogen related in any way to bacterial pyrogen or is it a totally independent substance? In other words, is a single endogenous pyrogen released from leucocytes of a given species in exactly the same way regardless of the bacterial pyrogen which releases it?

(2) Is the action of bacterial pyrogen simply to release pre-formed endogenous pyrogen, or does it accelerate its manufacture by leucocytes?

(3) Approximately how many molecules of endogenous pyrogen are needed to produce a recognizable fever? If it is comparatively few, as I suspect, then presumably these molecules are preferentially taken up by the central nervous system. But are they also preferentially taken up by temperature-regulating cells? And what effect would they have on the metabolism of such cells, because presumably it is on a metabolic change that the changed setting of the temperature mechanism depends?

I hope and expect that I shall leave this conference having had my ideas expanded and sharpened and perhaps with the answers to these three questions, and others.

REFERENCES

Bennett, I. L., and Beeson, P. B. (1953). *J. exp. Med.*, **98,** 477, 493.
Cooper, K. E., Cranston, W. I., and Honour, A. J. (1966). *J. Physiol., Lond.*, **186,** 22P.
Cooper, K. E., Cranston, W. I., and Snell, E. S. (1964). *Clin. Sci.*, **27,** 345.
Downey, J. A., Mottram, R. F., and Pickering, G. W. (1964). *J. Physiol., Lond.*, **170,** 415.
Gerbrandy, J., Cranston, W. I., and Snell, E. S. (1954). *Clin. Sci.*, **13,** 453.
Lewis, T., Pickering, G. W., and Rothschild, P. (1931). *Heart*, **15,** 359.
Liebermeister, C. von (1875). *Handbuch der Pathologie und Therapie des Fiebers.* Leipzig: Vogel.
Pickering, G. W. (1932). *Heart*, **16,** 115.
Snell, E. S. (1954). *J. Physiol., Lond.*, **125,** 361.
Stern, R. (1892). *Z. klin. Med.*, **20,** 63.
Ström, G. (1950). *Acta physiol. scand.*, **20,** 47, 97, 83.

SOME PHYSIOLOGICAL AND CLINICAL ASPECTS OF PYROGENS

K. E. Cooper

Division of Medical Physiology, University of Calgary, Calgary, Alberta

It would be a pity to open a symposium on fever without referring to Dr James Currie, the late 18th-century physician who was responsible for introducing the clinical thermometer into medical practice in England, and who studied many of the problems of the mechanism of fever. He investigated the epidemiology of some of the worst fevers of his time, besides doing the earliest known experimental work on accidental hypothermia. From Currie's time onwards much effort was expended in characterizing the patterns of response of body temperature to different types of infection or different phases of fever, until von Liebermeister (1875), again taking an experimental approach and studying the metabolic response to fever, produced evidence that now supports the view that fever represented a re-setting of the body's temperature regulating mechanisms at a new high level. Studies were made in 1911 by Jules Lefèvre and later by Eugéne du Bois (1936) on the metabolic responses of the whole body which occur during fever. The emphasis at this time was, quite reasonably, on attempting to cure the underlying cause of fever or to reduce the high body temperature by the empirical use of antipyretics. The real breakthrough came with the discovery by Grant and Whalen (1953) and Bennett and Beeson (1953) that a fever-producing substance could be obtained from rabbit white blood cells with or without their ever having been in contact with bacterial material, and with the work of Gerbrandy, Cranston and Snell (1954) in demonstrating that a new pyrogenic substance was liberated when human white blood cells were stimulated by bacterial pyrogen, and also with the work of Westphal and Lüderitz (1954) on the extraction and characterization of the highly pyrogenic lipopolysaccharide material obtainable from gram-negative organisms. The subsequent most important developments in Dr Barry Wood's laboratory and in the laboratories of Dr Atkins and his colleagues will become evident later in this symposium.

The story at present is that certain bacterial products, namely high molecular weight lipopolysaccharides, interact with white blood cells and stimulate them to release a substance which we now call leucocyte pyrogen.

Atkins and Snell (1964) have demonstrated that tissues other than leucocytes can have pyrogenic material extracted from them. This pyrogenic material has similar properties to leucocyte pyrogen, and this may explain the fevers that accompany aseptic tissue damage and allergic responses. Fever of this origin may, following organ transplantation, indicate tissue rejection as well as infection. Also some steroids may stimulate leucocytes to liberate pyrogen (Bodel and Dillard, 1966).

The terms "bacterial pyrogen" and "endotoxin" are both used to describe the pyrogenic material obtainable from microorganisms; however in this paper I shall stick to the term "bacterial pyrogen". Similarly the term "endogenous" pyrogen is frequently used to refer to the material derived from leucocytes, though it seems that the term "leucocyte pyrogen" is more descriptive and apt for the purpose. Pyrogens coming from other tissues could be called "tissue pyrogen", or more specifically a term could be added to denote their derivation.

Leucocyte pyrogen has been shown (Cooper, Cranston and Honour, 1967; Jackson, 1967; Repin and Kratzkin, 1967) in both the rabbit and the cat to act in minute quantities in the pre-optic area and the anterior hypothalamus close to the wall of the third ventricle and near the floor of the brain. Recently Rosendorff, Mooney and Long (1970) found evidence that there may also be a site of pyrogen action further back in the midbrain in the rabbit. Whether or not the leucocyte action on the brain involves monoaminergic pathways, and the extent to which it involves brain cations, will be discussed later in this symposium.

In man, Cooper, Johnson and Spalding (1964) have shown that leucocyte pyrogen does not appear to act on the nervous tissue in the spinal cord below the level of the sixth cervical vertebra or on the autonomic ganglia. The typical responses to pyrogen, namely peripheral vasoconstriction, shivering over the whole muscle mass of the body, and headache, do not occur in patients with a high spinal cord transection. Shivering does occur in a few muscles innervated from above the level of the transection, and if the transection is above the level of outflow of the sympathetic nerves no vascoconstriction is detectable on the body surface and, interestingly, the patient does not get headache.

Cooper, Cranston and Fessler (1960*a*, *b*) showed that leucocyte pyrogen can be precipitated or co-precipitated with ammonium sulphate and recovered; it is destroyed by precipitation with trichloroacetic acid and acetone, and inactivated by trypsin. It appears to be destroyed rapidly on the alkaline side of neutrality, particularly at pH values above 8·1, and more gradually at pH values below 6·5. Attempts were made to purify the leucocyte pyrogen by serial elution with different buffer molarities from DEAE columns. The best purification obtained was one in which a fever response of approximately 1°C rise in body temperature was produced by a fraction containing 4·75 μg protein. [Further purifications have been carried out (Rafter *et al.*, 1966; Gander, Mitchell and Goodale, 1970) and the leucocyte pyrogen identified as a lipid-polypeptide complex.] This seemed obviously impure and Dr Murphy will show how he has been able to purify this material more than 1000-fold. During purification we lost a large quantity of the purified material on storage in glass; comparison of the light absorption curves of the material before and after it lost its pyrogenicity showed that the material had lost an absorption peak at about 580 mμ. I mention this in the hope that it might be of use to someone attempting chemical purification, though at that wavelength it could also be the beginning of a red herring.

Most of the bacterial pyrogen is cleared from the circulation by a short rapid phase of a few minutes, and then the remainder by a longer phase (Rowley, Howard and Jenkin, 1956; Braude, Zalesky and

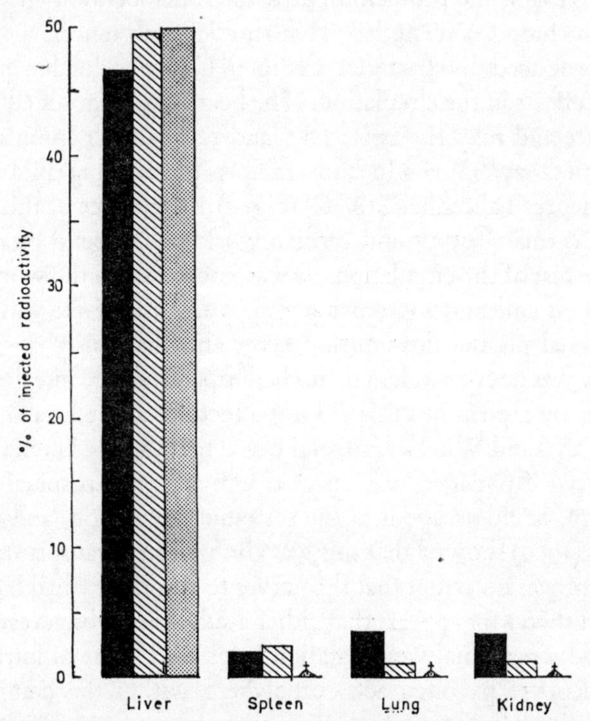

FIG. 1. Proportion of radioactive labelled bacterial pyrogen sequestered in various organs. Results of three independent observers. ■: [^{131}I] Pyrexal (Cooper and Cranston, 1963); ▧ : [^{51}Cr] *E. coli* endotoxin (Braude, Zalesky and Douglas, 1958); ▩: [^{32}P] *E. coli* endotoxin (Rowley, Howard and Jenkin, 1956).

Douglas, 1958; Cooper and Cranston, 1963). Within 30 minutes of injection of radioactive labelled bacterial pyrogen between 45 per cent and 50 per cent of the injected radioactivity was found in the liver, and some was also found in the spleen, the lung and the kidney (Fig. 1). The concentration of the radioactivity in the liver, the lung and the spleen rose well above that in the plasma, and small blood vessels in the lung could be seen packed with white blood cells and possibly platelets. It would be useful to investigate the effect of these sequestered cells upon the local blood flow and possibly upon the gas transport across alveoli of the lung. Their effect could be one of physical blocking; if large numbers of platelets are sequestered in these blood vessels it is likely that 5-hydroxytryptamine could be released locally (e.g. Henson, 1970), and possibly also kinins. Within seconds of entering the bloodstream bacterial pyrogen is to be found attached to both platelets and white blood cells, and it is tempting to think that these two blood elements may be the ferry-boats on which the injected noxious material is transported to the reticuloendothelial cells. But there remains the problem of how the reticuloendothelial cells recognize and trap blood elements laden with noxious substances.

Fever produced by bacterial pyrogen has considerable primary and secondary effects in the circulation. The heart rate is modified by the body temperature and may rise between 7 and 14 beats per minute per degree Fahrenheit (12–27/°C) rise in body temperature, giving an average of 10 beats per degree Fahrenheit (18/°C) (Fig. 2). The effect of this rise in heart rate upon cardiac output and cardiac work will depend on the physical state of the rest of the circulation. It was known from the work of Homer Smith and his colleagues (Goldring *et al.*, 1941) that there was a rise in the effective renal plasma flow during fever and that this change in kidney blood flow was not dependent on body temperature because the fever could be blocked by aspirin but this did not affect the altered renal blood flow. Cranston, Vial and Wheeler (1959) showed that the rise in renal blood flow also occurred if pyrogen was injected into the cerebrospinal fluid where little, if any, could escape into the systemic circulation, and Cooper and colleagues (1960) showed that this was also true in a transplanted kidney in which there was no doubt that the nerves to the kidney had been properly severed. It therefore appears that either a substance which reaches the kidney and induces a renal vasodilatation is released from an intracranial site, or the leucocyte pyrogen acts somewhere within the cranium causing the stimulation of nerves which cause the release of a humoral agent from some extra-cranial site. This is currently being investigated and as yet no clear picture has emerged. The renal vasodilatation has been measured by clearance of either iodopyracet (diodrast) or para-aminohippurate

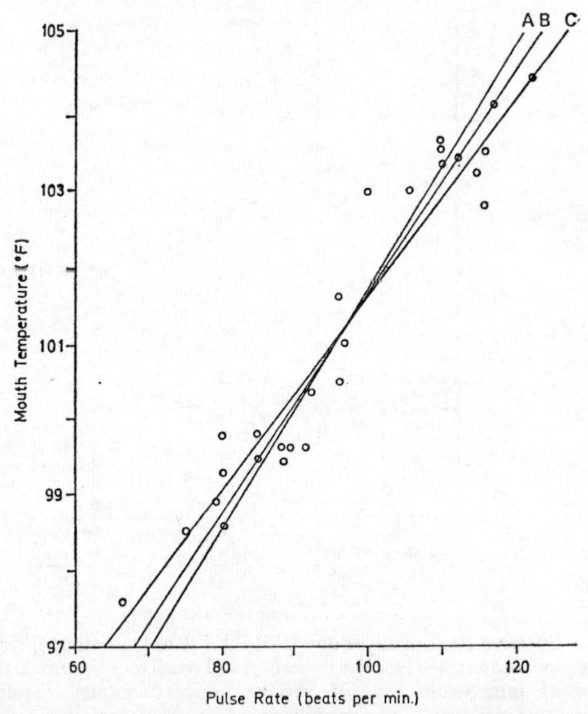

Fig. 2. Correlation between mouth temperature, T, and pulse rate, P, in a young man afflicted with a virus pneumonia. B = regression line; P = 96·33 + 7·13 (T − 101·16); A and C = 95% confidence limits of slope of B.

(PAH). Mrs Marty Houge and I (1968, unpublished) did some experiments in which the renal blood flow was measured directly by placing an electromagnetic flowmeter round the renal artery of the dog. Here the administration of intravenous typhoid vaccine caused a rise in renal blood flow which truly reflected a fall in renal vascular resistance under Nembutal anaesthesia. It occurred whether or not the anaesthesia was sufficiently light to allow the dog's temperature to rise due to the injected typhoid vaccine (Figs. 3 and 4). In this experimental situation it was usually found that there was an increase in urine flow during the renal vasodilatation and that the sodium concentration in the urine changed little. This implied a considerable loss of sodium in response to the intravenous vaccine. The renal blood flow, calculated from PAH clearances and arterial haematocrits, lay within 75–90 per cent of the measured renal artery flow even during the induced renal vasodilatation. Bradley and Conan (1947) have shown increases in hepatic and total splanchnic blood flow during fever. If the renal blood flow is of the order of one-fifth to one quarter of the cardiac output, then an increase in this, and in the liver blood flow, of some 40–70 per cent

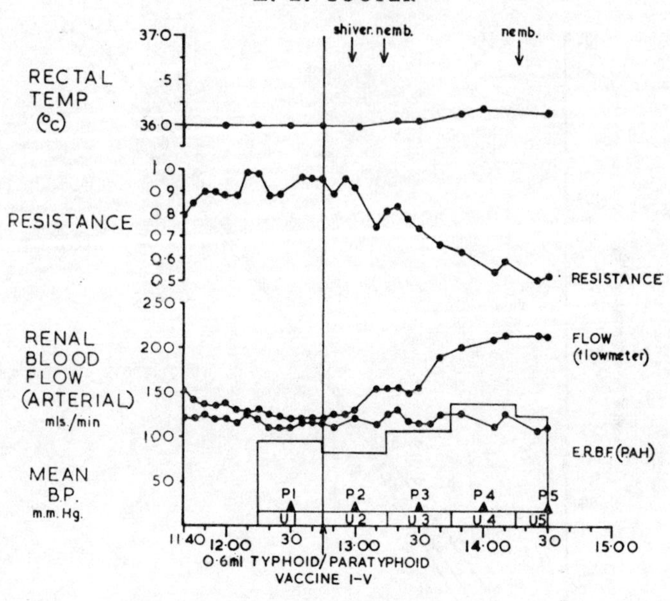

FIG. 3. Effective renal plasma flows (E.R.P.F.) (PAH clearances), renal artery blood flow, renal vascular resistance, and rectal temperature in the dog after intravenous typhoid vaccine under Nembutal (nemb.) anaesthesia.

will mean that a considerable increase in volume will be demanded of the heart's pumping capacity. This could be of some consequence in a patient whose heart was failing in its output.

Sudden flooding of the circulation with bacterial pyrogen can lead to responses which vary from an ordinary high fever to what is frequently called endotoxin shock. Schofield and co-workers (1968) studied the response that occurs when the spirochaetal type of louse-borne relapsing fever (*Borrelia recurrentis*) is treated by an injection of an antibiotic. The response to the antibiotic occurred about 60–90 minutes after its administration, and consisted of a rise in body temperature accompanied by a marked fall in the number of circulating leucocytes. Heart rate and pulse pressure also rose, but subsequently fell so that the main arterial pressure became very low and there was a considerable rise in respiratory rate. It is tempting to suggest that the rise in body temperature and the fall in leucocyte count were associated with the death of the organisms and the release of bacterial products such as bacterial pyrogens. The change in respiratory rate might also be associated with the sequestration of large numbers of leucocytes and platelets in the small vessels of the pulmonary circulation. But in view of the change in pulse pressure this change in respiratory rate might be more readily explained if those sequestered

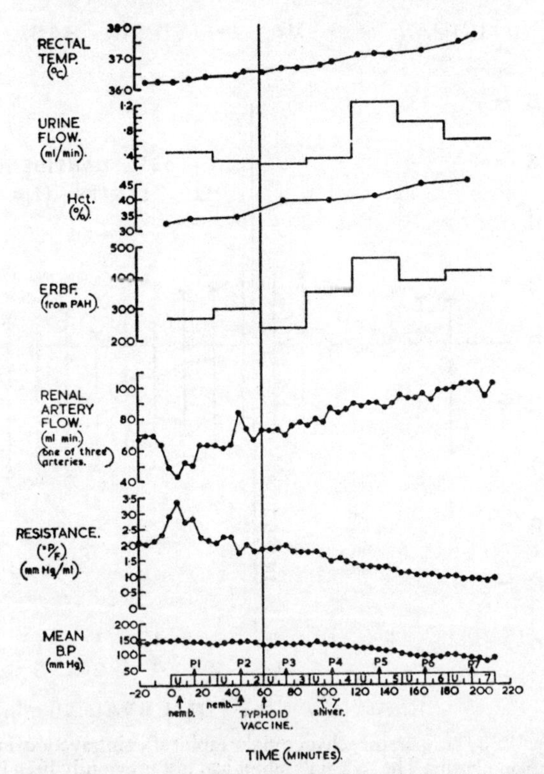

FIG. 4. Effective renal blood flows (E.R.B.F.) [from PAH clearances and haematocrit (Hct)], renal artery flows, vascular resistance (measured in one of three renal arteries), rectal temperature, and urine volume in the dog after intravenous typhoid vaccine under Nembutal (nemb.) anaesthesia.

leucocytes and platelets released some substance which stimulated respiration rather than actually plugging up the vessels. The evidence available at present has not demonstrated the presence of leucocyte pyrogen during this reaction, but this may only be due to the quantity of blood used for its detection or the time at which blood samples were taken.

Clearly to solve such problems it is necessary to be able to detect and if possible assay human leucocyte pyrogen and circulating bacterial pyrogen in human blood. Mr D. A. MacLachlan and I have evidence that it is possible to do both using the method described by Murphy (1967) for rabbit blood. Human plasma can be injected into rabbits which have never been given human plasma intravenously, without their getting fever, provided that the quantity injected is kept low (Fig. 5). Human blood to which bacterial pyrogen had been added in the cold was centrifuged and the plasma mixed with an equal volume of rabbit plasma. This was then injected into rabbits which had never been injected with human plasma,

HUMAN PLASMA → VIRGIN RABBITS

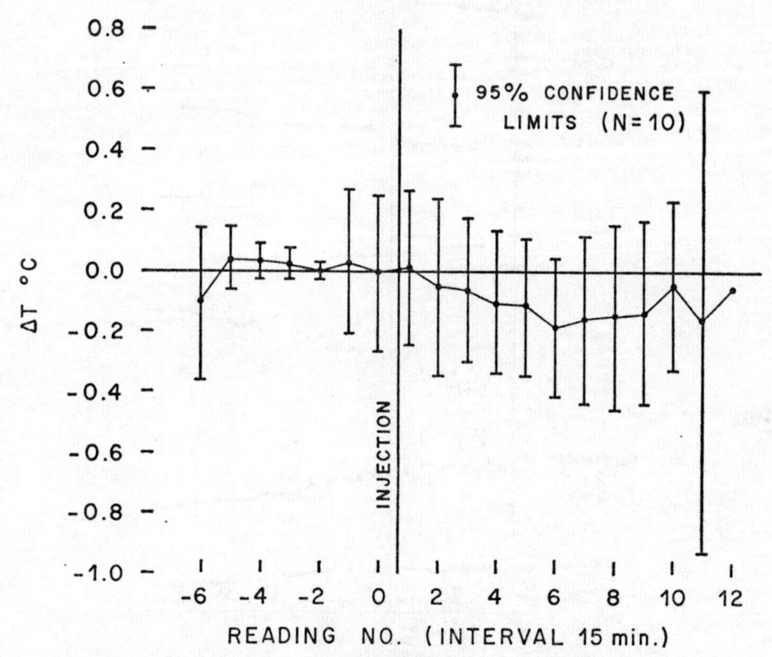

Fig. 5. Body temperature change in the rabbit after intravenous injection of human plasma. The "virgin" rabbit had not previously been injected with human plasma.

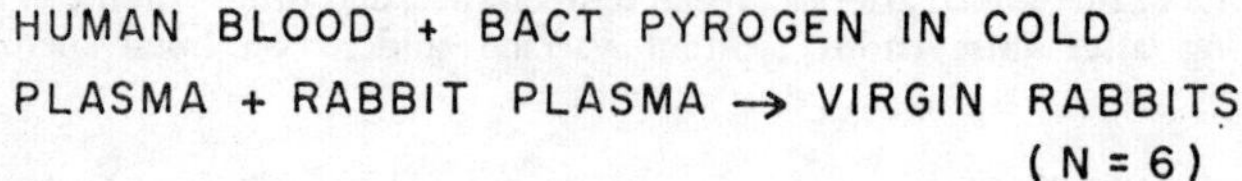

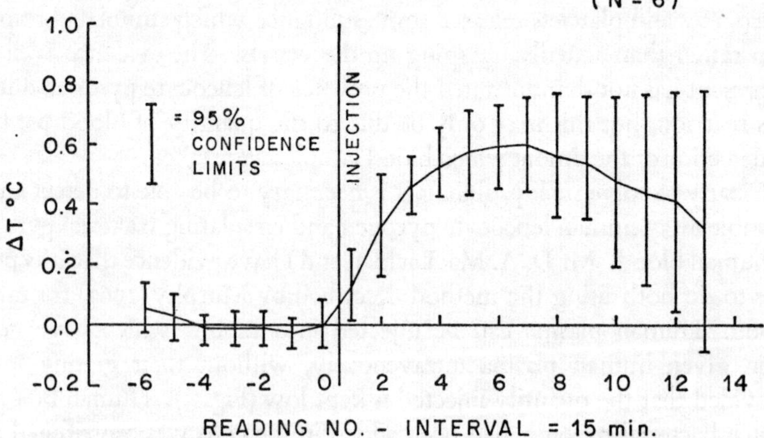

Fig. 6. Temperature change when human plasma containing bacterial pyrogen and rabbit plasma was injected intravenously into rabbits which had not previously received human plasma ("virgin rabbits").

and these rabbits all got fever (Fig. 6). Human blood was then incubated with the same dose of bacterial pyrogen, centrifuged, and the plasma added to an equal volume of rabbit plasma. The mixture was then incubated for a further 12 hours after which it caused fever when injected into rabbits rendered refractory to bacterial pyrogen by a large dose on the previous day (Fig. 7). Human plasma to which bacterial pyrogen had been added in the cold and then incubated in an equal volume of rabbit plasma did not cause fever when injected in equivalent amounts into refractory rabbits

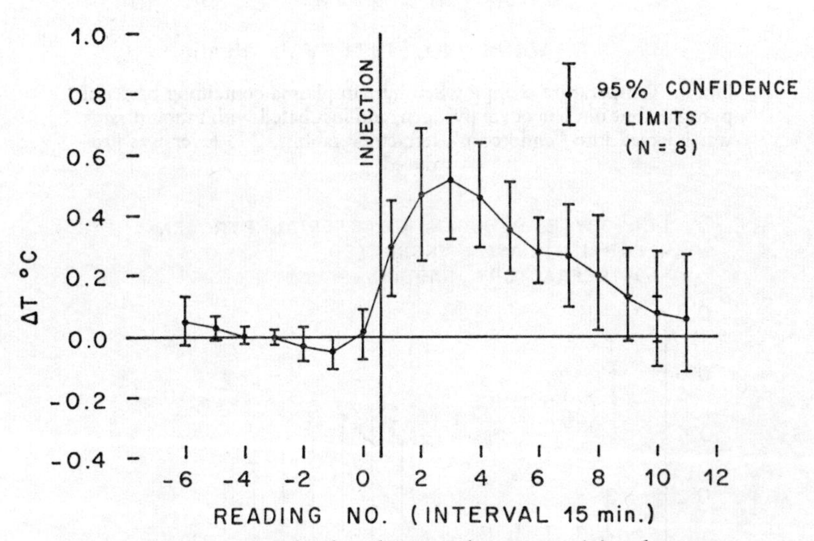

FIG. 7. Fever produced when human plasma containing leucocyte pyrogen and incubated with rabbit plasma was injected into "endotoxin"-refractory rabbits.

(Fig. 8). Human blood incubated with bacterial pyrogen was centrifuged and an equal volume of rabbit plasma added. Another dose of bacterial pyrogen was added to this mixture and then it was divided into two parts, A and B. Part A was incubated for 12 hours and then injected into the refractory rabbits. A slight fever in approximately the same range as in Fig. 5 was produced (Fig. 9). Part B was injected into rabbits which were not refractory and a much higher and more prolonged fever was obtained (Fig. 10). It thus appears that leucocyte pyrogens and bacterial pyrogen can be detected in human blood by a rabbit assay. It also appears that leucocyte pyrogen can be detected in human blood in the presence of a

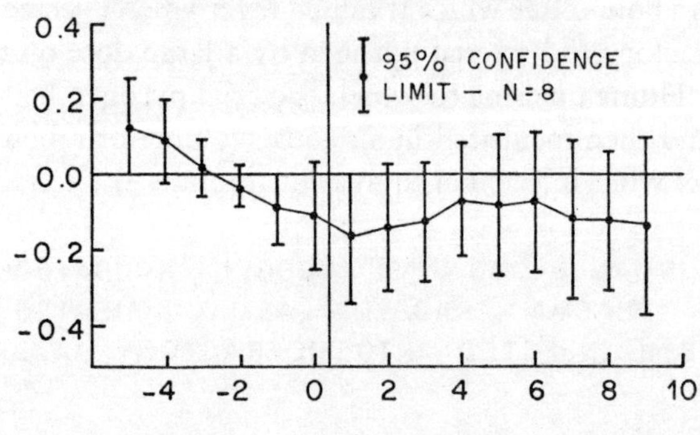

FIG. 8. Temperature change when human plasma containing bacterial pyrogen, but not leucocyte pyrogen, and incubated with rabbit plasma was injected into "endotoxin"-refractory rabbits. No fever was produced.

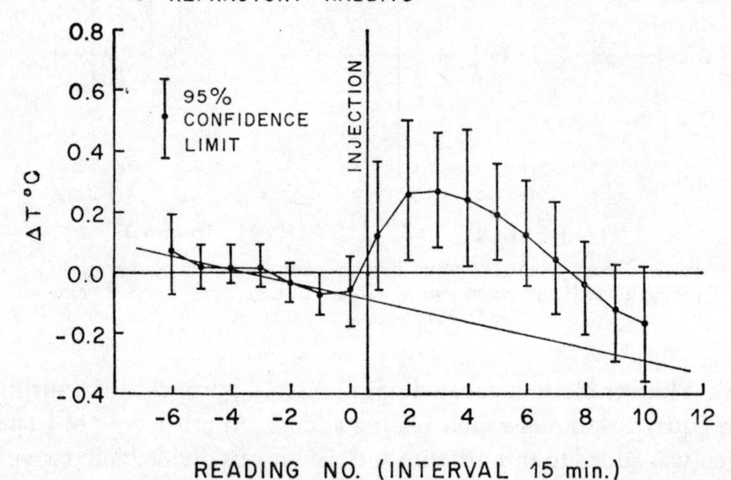

FIG. 9. Rabbit fever response (ΔT°C) to a mixture of human leucocyte pyrogen and bacterial pyrogen which had been incubated with rabbit plasma and injected into "endotoxin"-refractory rabbits. Eight rabbits were used.

considerable amount of bacterial pyrogen. A quantitative assay is more difficult since it is preferable to use a cross-over technique to make the analysis truly quantitative. Fig. 11 shows that quite a sharp fever occurs if even

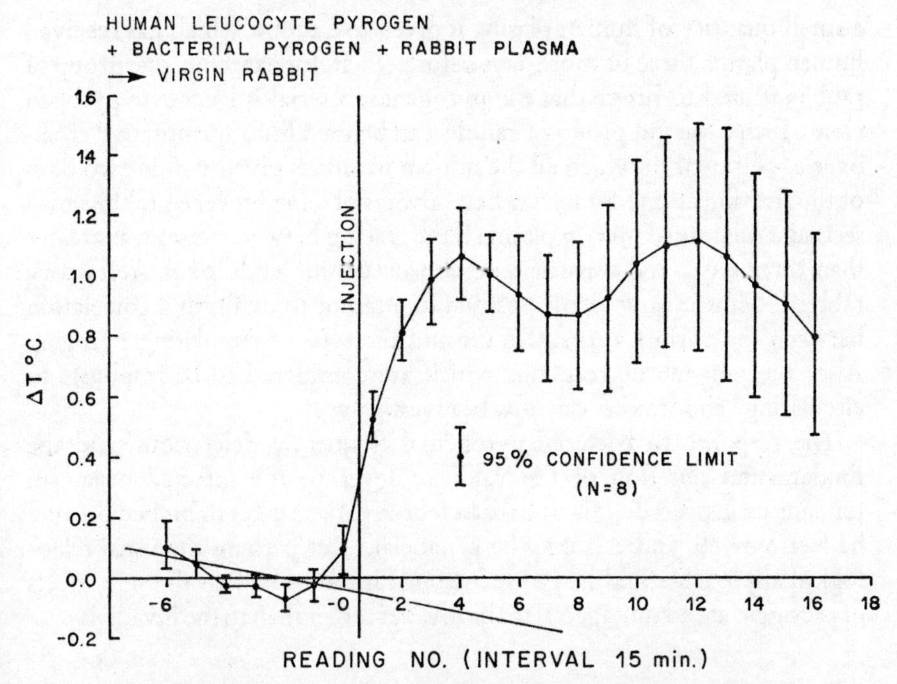

FIG. 10. Rabbit fever response (ΔT°C) to a mixture of human leucocyte pyrogen, bacterial pyrogen and rabbit plasma, injected into "virgin rabbits".

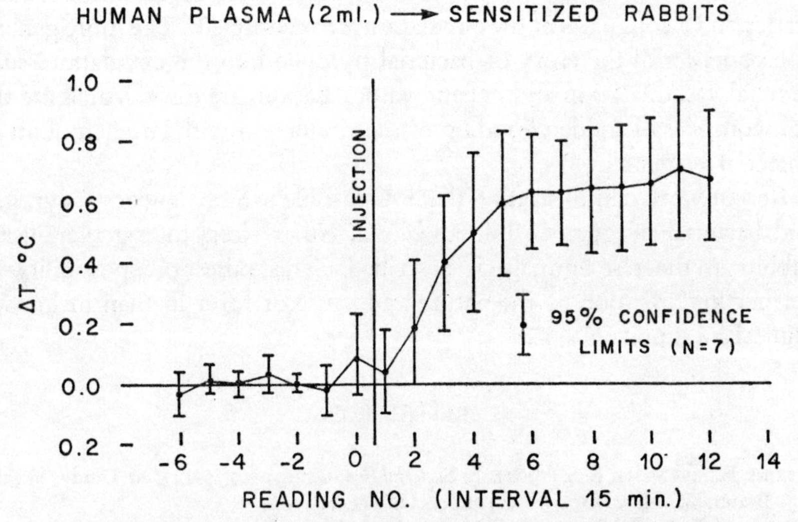

FIG. 11. Fever produced by injecting 2 ml human plasma into rabbits which had received a similar injection 3–6 days previously.

a small quantity of human plasma is given to a rabbit which has received human plasma three or more days before. So if, for example, one group of rabbits is used to prove that the pyrogenic material is leucocyte pyrogen alone, then a second group of rabbits can be used for a quantitative cross-over experiment in which all the human plasma is given within two days of the first injection. So far we have always obtained fever in rabbits by a second 2 ml dose of human plasma if the spacing between doses was greater than three days, and this response also occurs in "endotoxin"-refractory rabbits. Thus it now seems possible to attempt to establish a correlation between the various types of fever and the type of circulating pyrogen. Also, the non-febrile reactions which are considered to be responses to circulating "endotoxin" can now be investigated.

The response to bacterial pyrogen is frequently deleterious, and the fundamental question of the value of fever to the infected organism remains unanswered. It is true that a febrile patient takes to his bed because he feels unwell, and this may be beneficial. But perhaps the usual teleological approach could be twisted round; it might be that the side effects of pyrogens are advantageous to the invader rather than to the invaded.

SUMMARY

A plea is made for greater standardization in the terminology of pyrogens. Early attempts to purify leucocyte pyrogen and some work on the clearance of bacterial pyrogen from the circulation are described, and the implications of the attachment of pyrogen to blood-formed elements and their sequestration in various parts of the circulation are mentioned. The more general consequences of the entry of bacterial pyrogen into the circulation, such as renal vasodilatation and hepatic vasodilatation, are discussed, as are the consequences of sudden flooding of the circulation with large amounts of bacterial pyrogen.

Recent work demonstrating that it is possible to assay leucocyte pyrogen and bacterial pyrogen in human blood, either alone or together, using rabbits as the test animals, is described. This raises the possibility of further investigation of the nature and cause of fever in man in known clinical conditions.

REFERENCES

Atkins, E., and Snell, E. S. (1964). In *Bacterial Endotoxins*, pp. 134–143, ed. Landy, M., and Braun, W. New Brunswick: Rutgers University Press.

Bennett, I. L., Jr, and Beeson, P. B. (1953). *J. exp. Med.*, **98,** 493–508.

Bodel, P. T., and Dillard, M. (1966). *J. clin. Invest.*, **45,** 988.

BRADLEY, S. E., and CONAN, N. J. (1947). *J. clin. Invest.*, **26**, 1175.
BRAUDE, A. I., ZALESKY, M., and DOUGLAS, H. (1958). *J. clin. Invest.*, **37**, 880.
COOPER, K. E., and CRANSTON, W. I. (1963). *J. Physiol., Lond.*, **166**, 41–42P.
COOPER, K. E., CRANSTON, W. I., DEMPSTER, W. J., and MOTTRAM, R. F. (1960). *J. Physiol., Lond.*, **155**, 21–22P.
COOPER, K. E., CRANSTON, W. I., and FESSLER, J. H. (1960*a*). *J. Physiol., Lond.*, **152**, 51–52P.
COOPER, K. E., CRANSTON, W. I., and FESSLER, J. H. (1960*b*). *J. Physiol., Lond.*, **154**, 22–23P.
COOPER, K. E., CRANSTON, W. I., and HONOUR, A. J. (1967). *J. Physiol., Lond.*, **191**, 325–337.
COOPER, K. E., JOHNSON, R. H., and SPALDING, J. M. K. (1964). *J. Physiol., Lond.*, **171**, 55P.
CRANSTON, W. I., VIAL, S. V., and WHEELER, H. O. (1959). *Clin. Sci.*, **18**, 579.
DU BOIS, E. F. (1936). *Basal Metabolism in Health and Disease.* Philadelphia: Lea and Febiger.
GANDER, G. W., MITCHELL, R., and GOODALE, F. (1970). *Fedn Proc. Fedn Am. Socs exp. Biol.*, **29**, 690.
GERBRANDY, W., CRANSTON, W. I., and SNELL, E. S. (1954). *Clin. Sci.*, **13**, 453–459.
GOLDRING, W., CHASIS, H., RANGES, H. A., and SMITH, H. W. (1941). *J. clin. Invest.*, **20**, 637.
GRANT, R., and WHALEN, W. J. (1953). *Am. J. Physiol.*, **173**, 47.
HENSON, P. M. (1970). *J. exp. Med.*, **131**, 287–306.
JACKSON, D. L. (1967). *J. Neurophysiol.*, **30**, 586.
LEFÈVRE, J. (1911). *Chaleur animale et bioénergétique.* Paris: Masson.
LIEBERMEISTER, C. VON (1875). *Handbuch der Pathologie und Therapie des Fiebers.* Leipzig: Vogel.
MURPHY, P. A. (1967). *J. exp. Med.*, **126**, 745.
RAFTER, G. W., CHEUK, S. F., KRAUSE, D. W., and WOOD, B. W. (1966). *J. exp. Med.*, **123**, 433–444.
REPIN, I. S., and KRATZKIN, I. L. (1967). *Sechenov physiol. J. USSR.*, **50**, 1206–1211.
ROSENDORFF, C., MOONEY, J. J., and LONG, C. N. H. (1970). *Fedn Proc. Fedn Am. Socs exp. Biol.*, **29**, 523, abst. 1547.
ROWLEY, D., HOWARD, J. G., and JENKIN, C. R. (1956). *Lancet*, **1**, 366–367.
SCHOFIELD, T. P. C., TALBOT, J. M., BRYCESON, A. D. M., and PARRY, E. H. O. (1968). *Lancet*, **1**, 58–62.
WESTPHAL, O., and LÜDERITZ, O. (1954). *Angew. Chem.*, **66**, 407–517.

DISCUSSION

Pickering: What is the difference between refractoriness and tolerance?

Cooper: I have always used the term tolerance to refer to the diminishing response in animals which have had a series of doses of bacterial pyrogen over several days. These animals do become refractory, but in the sense that I am using this term, which is how Murphy used it, refractoriness refers to animals which had one big dose of bacterial pyrogen the day before the test was applied. Their response to bacterial pyrogen was grossly diminished. I don't know if there is any difference in mechanism between refractoriness and tolerance.

Atkins: An animal which has had a single large dose of exogenous pyrogen (such as bacterial endotoxin) 24 hours earlier may show a diminished febrile response when an unrelated pyrogen is given. Since such an animal responds normally to injected endogenous pyrogen it appears to be temporarily unable to *produce* endogenous pyrogen when given the second agent. Dr Snell and I have referred to such an animal as "refractory" in contrast to the animal made conventionally tolerant by daily injections of endotoxin (Snell and Atkins, 1965). Tolerant animals respond normally to unrelated pyrogens and unlike refractory animals appear to have no impairment in their ability to mobilize endogenous pyrogen to other stimuli. Their failure to respond normally to the tolerance-inducing agent appears to be specific and is at least partly due to the development of "blocking" antibodies since this form of tolerance can be passively transferred to normal recipients with serum (Greisman, Carozza and Hills, 1963). It is clear that immunological mechanisms play an important role in tolerance, presumably by inactivating the pyrogenic stimulus so that it has less effect on the target cell that produces endogenous pyrogen.

Bondy: Simple exhaustion of the cells which produce endogenous pyrogen seems unlikely since daily spikes of fever can be produced by giving large doses of steroid pyrogen day after day (Kappas, Glickman and Palmer, 1960). Steroid pyrogens do not produce tolerance even after many doses, so the system may be quite different. But if one believes that the steroid system produces a leucocyte pyrogen analogous to that produced in response to a bacterial pyrogen, then exhaustion of the leucocyte pyrogen-producing system cannot be the explanation.

Cooper: I think we have part of the answer to this. We injected rabbits which had previously received human plasma intravenously with a large dose of bacterial pyrogen. The following day, when they were refractory to bacterial pyrogen, we gave them more human plasma. They still developed fever. Therefore it appears that some tissue was capable of mobilizing an endogenous type of pyrogen.

Pickering: I would have thought that the simplest thing to do would be to prepare leucocytes from an animal and see how much endogenous pyrogen could be obtained by the appropriate stimulus, say bacterial pyrogen, then make the animal refractory, collect the leucocytes again and see if you get the same amount of endogenous pyrogen or less.

Atkins: We have done that using Kupffer cells of the liver (Dinarello, C. A., Bodel, P. T., and Atkins, E., unpublished). If rabbits are given a single large injection of bacterial pyrogen (endotoxin) and the Kupffer cells are taken out the next day, these cells are less reactive to either the same or an unrelated pyrogenic stimulus. This is clear evidence that there is a form of refractoriness in which the cell is unable to mobilize normal amounts of pyrogen regardless of the stimulus. However, since *in vitro* refractoriness can also be demonstrated when normal Kupffer cells are incubated in the serum of an animal that has received a large dose of endotoxin, an immunological mechanism also seems to contribute to this form of reduced responsiveness, as it does in conventional tolerance.

Pickering: If it is true about Kupffer cells, is it also true about leucocytes, or do you now think that the leucocytes are comparatively unimportant in producing endogenous pyrogen and in the mechanism of fever?

Atkins: There is little doubt that circulating leucocytes play a major role in endotoxin-induced fever.

Pickering: What happens to leucocytes in refractoriness?

Bodel: Blood leucocytes from animals that have received a single large dose of endotoxin (typhoid vaccine) the preceding day release less leucocyte pyrogen when incubated with endotoxin *in vitro*, but respond normally to a heterologous stimulus such as phagocytosis (Dinarello, C. A., Bodel, P. T., and Atkins, E., unpublished). However, the cells of animals that have become conventionally tolerant to typhoid vaccine through a course of daily injections respond normally to endotoxin as well as to heterologous stimuli (Collins and Wood, 1959; Dinarello, Bodel and Atkins, 1968).

Whittet: It takes quite a long time to get an animal thoroughly tolerant. In some of my experiments on rabbits it took at least a month before the secondary peak of fever disappeared after a daily large dose of bacterial pyrogen. You can abolish the tolerance immediately by giving thorium dioxide, so the reticuloendothelial system also plays quite an important part (Whittet, 1967).

Cranston: The other kind of refractoriness, described by Greisman and co-workers (1967), may or may not be similar to the refractoriness induced by a single large dose of endotoxin. Greisman and his colleagues showed that a continuous infusion of bacterial pyrogen causes an initial febrile response which lasts for only a few hours. The temperature returns to baseline despite continuation of the infusion.

Atkins: Greisman has interpreted this finding in terms of a specific desensitization of the cell to the infused endotoxin since his animals were fully responsive if given a heterologous stimulus such as tuberculin (Greisman and Woodward, 1965). These experiments also clearly show that tolerance is not due to an inability of the thermoregulatory centre to respond to a pyrogenic stimulus.

Cranston: But that implies that they were sensitized beforehand.

Atkins: Many people believe that this is one of the mechanisms of action of endotoxin.

Pickering: Do you mean that a bacterial pyrogen produces fever by a sort of mechanism of hypersensitivity?

Atkins: I believe that Greisman's experiments are consistent with that theory: that is, a continuous infusion with endotoxin will desensitize animals so that they will no longer respond to this agent, but will respond normally to another pyrogen.

Snell: I am not sure that by desensitization Greisman meant the same mechanism as desensitizes the classical sensitivity reaction. He meant that by a continuous infusion something happens at the cellular level to stop endotoxin exerting its usual action.

Landy: We have abundant evidence that as judged by the presence of natural antibodies (against bacteria and foodstuffs) in their serum (Michael, Whitby and Landy, 1962) all mammalian hosts have been in contact with endotoxic materials or their derivatives and are sensitized.

Pickering: Isn't it now possible to rear animals in entirely germ-free conditions?

Landy: Yes. However, germ-free rodents show the same antibody reactivities and responsiveness to endotoxin as conventional animals, but at a lower level. Only ungulates, deprived of any contact with colostrum, show what can be regarded as an immunologically virgin situation.

Bondy: In connection with the Homer Smith experiments (Goldring *et al.*, 1941), it is worth commenting that increased renal vasodilatation can be observed even if the fever response is blocked after administration of bacterial pyrogen. Does this mean that the response is a direct effect of the pyrogen (or of the leucocyte pyrogen it may produce) on the kidney?

Cranston: Intravenous injections of bacterial or endogenous pyrogen cause renal vasodilatation, and the rise in temperature is reasonably correlated with the increase in effective renal plasma flow (Cranston, Vial and Wheeler, 1969). Intrathecal injection of bacterial pyrogen also causes renal vasodilatation, and the relationship between temperature rise and increased plasma flow is similar to that after intravenous injection. In this situation there is no detectable circulating leucocyte pyrogen, unlike the situation after intravenous injections. This makes it unlikely that either bacterial or leucocyte pyrogen directly affects the renal vasculature. It is unlikely that the vasodilatation is neurally mediated, because it still occurs in an auto-transplanted kidney (Cooper *et al.*, 1960).

Landy: Dr Cooper, in the tissue and organ distribution studies you described, large doses of bacterial endotoxin were used. Some of us would regard these amounts as rather unphysiological. Therefore what significance can you attach to the subsequent experiments when the dose of pyrogen was in contrast a very modest one?

Cooper: Large doses of endotoxin were used so that enough radioactivity to determine the distribution of the endotoxin in the animal was introduced;

another unphysiological aspect is that the animals were anaesthetized. So all one can say is that a large dose of bacterial pyrogen is distributed as we showed. This may one day prove to have some relevance to the normal mechanism of fever, or it may not.

Pickering: Your Fig. 1 (p. 7), Dr Cooper, only accounts for about 50 per cent of the radioactivity. Where did the rest go?

Cooper: Some went into the bile (in very small amounts), some was excreted by the kidney and there was some still left in the circulation. At the time we were able to account for nearly all of it.

Bondy: Did you look in the lung?

Cooper: In our experimental conditions it is certainly in the lungs in a greater concentration than in plasma; also there are quite a lot of sequestered cells in the lungs.

REFERENCES

COLLINS, R. D., and WOOD, W. B. (1959). *J. exp. Med.*, **110**, 1005.

COOPER, K. E., CRANSTON, W. I., DEMPSTER, W. J., and MOTTRAM, R. F. (1960). *J. Physiol., Lond.*, **155**, 21–22P.

CRANSTON, W. I., VIAL, S. U., and WHEELER, H. O. (1969). *Clin. Sci.*, **18**, 579–585.

DINARELLO, C. A., BODEL, P. T., and ATKINS, E. (1968). *Trans. Ass. Am. Physns*, **81**, 334.

GOLDRING, H., CHASIS, H., RANGES, H. A., and SMITH, H. W. (1941). *J. clin. Invest.*, **20**, 637–653.

GREISMAN, S. E., CAROZZA, F. A., and HILLS, J. D. (1963). *J. exp. Med.*, **117**, 663.

GREISMAN, S. E., HORNICH, R. B., WAGNER, H. N., and WOODWARD, T. E. (1967). *Trans. Ass. Am. Physns*, **80**, 150–158.

GREISMAN, S. E., and WOODWARD, W. E. (1965). *J. exp. Med.*, **121**, 911.

KAPPAS, A., GLICKMAN, P. B., and PALMER, R. H. (1960). *Trans. Ass. Am. Physns*, **73**, 176–185.

MICHAEL, J. G., WHITBY, J. L., and LANDY, M. (1962). *J. exp. Med.*, **115**, 131.

SNELL, E. S., and ATKINS, E. (1965). *J. exp. Med.*, **121**, 1019.

WHITTET, T. D. (1967). In *Scientiae Pharmaceuticae* (Proceedings of the 15th Congress of Pharmaceutical Sciences, Prague 1965), vol. 2, pp. 283–290, ed. Hanc, O., and Hubik, J. London: Butterworth.

PRODUCTION, CHEMISTRY AND PROPERTIES OF BACTERIAL PYROGENS AND ENDOTOXINS

ELIZABETH WORK

Biochemistry Department, Imperial College of Science & Technology, London

"ENDOTOXINS possess an intrinsic fascination that is nothing less than fabulous. They seem to have been endowed by Nature with virtues and vices in the exact and glamorous proportions needed to render them irresistible to any investigator who comes to know them. They intrigue the chemist. The molecular basis for their biological action seems always on the verge of discovery but somehow just eludes detection." (Bennett, 1964.)

The most active pyrogens of microbial origin are those produced by gram-negative bacteria. Siebert (1923, 1925) found that the pyrogenic substance which developed in freshly distilled water on standing in non-sterile conditions was produced by gram-negative bacteria. It was destroyed only slowly by heat, but was labile to hot acids or alkalis. Since this observation, the active pyrogen has been isolated and reasonably well characterized; it is a high-molecular weight, very complex polymer known as "endotoxin" and has very potent and diverse biological properties. The pyrogenic activity is extremely high in the rabbit, and intravenous doses of about 0·001 μg of endotoxin per kilogram body weight, or of 1000 killed bacterial cells, will produce a pyrogenic response. The toxic dose of endotoxin is some 10 000 times higher.

The main toxic and pyrogenic principals of endotoxins, the cell wall lipopolysaccharides, have been well reviewed (Lüderitz, Staub and Westphal, 1966; Lüderitz, Jann and Wheat, 1968; Nowotny, 1969; Lüderitz, 1970). They are potent antigens, responsible for O-antigenic specificities of Enterobacteriaceae. Endotoxins can be prepared from bacteria by a variety of methods which can determine their composition and biological activities. They all contain lipopolysaccharides in varying proportions. The literature in the field is often confused because the source, method of preparation, and nature of the product are not always specified, and the nomenclature is not standardized. Most endotoxin preparations probably consist of complexes or mixtures of proteins, phospholipids and lipopolysaccharides. It is proposed in this review to restrict the name "endotoxin" to such preparations from gram-negative bacteria, and not to extend it to separated lipopolysaccharides as has been frequently done in the past. (The basis for this distinction is partly historical, since the original "Boivin" preparations of endotoxins, also called O-somatic antigens, were these complexes.) The term lipopolysaccharide will refer to the lipopolysaccharide moiety largely, if not entirely, freed from proteins and phospholipids.

PREPARATION OF ENDOTOXINS AND LIPOPOLYSACCHARIDES

The main methods which have been used to obtain endotoxins from bacterial sources are given in Table I. These fall into several categories;

TABLE I

SOME METHODS OF PRODUCTION OF PYROGENS* FROM GRAM-NEGATIVE BACTERIAL CULTURES

(A) Cells

	Extractant	*Applicability*	*Reference*
(1)	Trichloroacetic acid	Only smooth forms	Boivin and Mesrobeanu, 1933
(2)	Diethylene glycol	Limited	Morgan, 1937
(3)	Aqueous pyridine	Limited	Goebel, Binkley and Perlman, 1945
(4)	Aqueous ether/NaCl	Wide	Ribi, Milner and Perrine, 1959; Ribi *et al.*, 1961
(5)	1M-NaCl : 0·1M-Na citrate	Unlimited	Raynauld, Digeon and Nauciel, 1964
(6)	EDTA	Wide	Leive 1965; Leive, Shovlin and Mergenhagen, 1968; Cox and Eagon, 1968
(7)	Aqueous phenol	Unlimited, except for very rough mutants	Westphal, Lüderitz and Bister, 1952
(8)	Phenol-chloroform-petroleum ether	Very rough mutants	Galanos, Lüderitz and Westphal, 1969

(B) Culture filtrates

Type of culture	*Reference*
Colicinogenic	Goebel and Barry, 1958
Lysine-limited *E. coli* 12408	Knox, Cullen and Work, 1967; Work, 1969
Chromobacterium violaceum	Corpe and Salton, 1966
High density, Enterobacteriaceae	Marsh and Crutchley, 1967; Crutchley, Marsh and Cameron, 1968

* Products are endotoxins except for *A*7 where it is lipopolysaccharide, and *A*8 where it is glycolipid.

method 1 (Boivin and Mesrobeanu, 1933) is only applicable to "smooth-type" organisms. Fresh cells are treated with 0·25 per cent trichloroacetic acid for three hours at 2°C. The solids are removed by centrifugation (4000 rev./min), and the supernatant fluid is poured slowly into two volumes of cold ethanol. After standing overnight at 2°C the precipitate is spun down and dialysed. When first studied by Boivin this preparation of somatic antigen was designated a "glycolipid", but it was subsequently shown by Morgan and Partridge (1940) to be a protein-polysaccharide-lipid complex. Similar types of complexes which were also shown to contain proteins, phospholipids (termed lipid B), polysaccharides and bound lipids (lipid A) were prepared using milder extractants (methods 2, 3, 4), the most widely applicable technique (method 4) being that introduced by Ribi and co-workers (1959, 1961). This involves suspending washed cells (smooth or rough strains) in 0·15 M-NaCl at 18°C and shaking for one minute with an excess of ether: after standing overnight the centrifuged, aqueous phase is dialysed and treated at 4°C with 0·15 M-NaCl and 68 per cent (v/v) ethanol. The precipitate so obtained ("crude endotoxin") represents about 5 per cent of the cell weight but is reported to contain less lipid A than the Boivin-type endotoxin (method 1). Another mild extraction method (5) devised by Raynaud, Digeon and Nauciel (1964) extracts cells with a hypertonic solution of 1 M-NaCl+0·1 M-Na citrate for one day at 0°C.

A completely different reagent (method 6) ethylene-diaminetetra-acetate (EDTA) was found by Leive (1965) to release lipopolysaccharide from *Escherichia coli* and other Enterobacteriaceae; subsequently Cox and Eagon (1968) used it for *Pseudomonas aeruginosa*. In both cases the lipopolysaccharide was complexed to protein and lipid (Leive, Shovlin and Mergenhagen, 1968). The method could be applied to walls or cells and may prove to be widely applicable. Washed cells or isolated walls are suspended at 25–37°C in 0·12 M-tris-HCl, pH 8, and treated for 4–30 minutes with EDTA, pH 8, at a concentration of 1 μmol/1·5 mg walls per millilitre, or 0·5 μmol/10 mg wet wt. (10^{10} cells) (Leive, Shovlin and Mergenhagen, 1968; Rogers, Gilleland and Eagon, 1969). About half of the total cell lipopolysaccharide is thus removed. Methods 7 and 8, originated by Westphal and co-workers, produce preparations of lipopolysaccharides relatively free from protein and phospholipids, but of equal biological potencies to the full endotoxins. Either whole cells, cell walls (Holme *et al.*, 1969), or endotoxin preparations (Knox, Cullen and Work, 1967) can be used, but less pure materials are obtained from whole cells. Method 7 is most widely applicable and is the standard method for preparation of lipopolysaccharides. It does not give satisfactory preparations of very lipophilic lipopolysaccharides such as are found in a heptose-less R_e mutant of *Salmonella minnesota* (see later), *Xanthomonas campestris*, or *Citrobacter* (Kasai and Nowotny, 1967). For these unusual strains the modification shown as method 8 can be used. In the standard method 7, the cells, cell walls or endotoxin are heated at 68°C for 5–20 minutes in a mixture of 90 per cent phenol and water (1 : 1 v/v). After cooling to 4°C, the top aqueous phase is separated by centrifugation and freed from phenol by extraction with ether or by dialysis. It is concentrated (if necessary) under reduced pressure and the lipopolysaccharide is sedimented by repeated centrifugation at 100 000**g** (yield 1–4 per cent of dry bacteria). Lipopolysaccharide can also be obtained or further purified by precipitation with ethanol (10 vol.) and then with 0·025 M-$MgCl_2$ (Osborn *et al.*, 1962; Taylor, Knox and Work, 1966).

The other methods described in Table I relate to typical endotoxins which can be isolated from filtrates of bacterial cultures, here referred to as "excreted" endotoxins to differentiate them from the "extracted" preparations already described. The term endotoxin, used originally to distinguish the toxin from bacterial exotoxins, is obviously now a misnomer. The original isolation of substantial amounts of extracellular lipopolysaccharide-containing materials was made by Goebel and Barry (1958) who found the typical components of endotoxins in filtrates of "colicinogenic" *E. coli* cultures. Bishop and Work (1965) isolated lipopolysaccharides from material in the culture filtrate of a lysine-requiring mutant of *E. coli* (12408) grown under lysine-limiting conditions with high aeration. The material was subsequently shown to contain protein (10 per cent), lipopolysaccharide (60 per cent) and phospholipid (25 per cent) (Knox, Cullen and Work, 1967). It could be isolated in high yields (up to 2 g per litre) by mixing the culture filtrate vigorously with chloroform which produced a micellar "precipitate" separable by centrifuging; after removing the chloroform *in vacuo* an opalescent turbid solution was obtained and dialysed (Knox, Cullen and Work, 1967; Janda and Work, unpublished). The other work cited in Table I has produced similar complexes, excreted by normal cultures of gram-negative bacteria, especially those grown to high cell densities (10·7 g/l) (Marsh and Crutchley, 1967).

TABLE II

PYROGENICITY OF FRACTIONS FROM *E. coli* O113*

Fraction	*Pyrogenicity* FI_{40}, µg
Cells, aqueous ether extract (endotoxin)	0·39
Cells, trichloroacetic acid extract (endotoxin)	1·6
Walls, whole	3·0
Protoplasm, whole	140
Walls, phenol extract (lipopolysaccharide)	0·16

* Ribi *et al.*, 1964

FI_{40}: Fever Index (Haskins *et al.*, 1961)

Dialysis or ammonium sulphate precipitation was used to obtain material from the culture filtrates.

LOCATION AND FUNCTION OF ENDOTOXINS IN BACTERIAL CELLS

The main location of endotoxins is in the cell walls (Table II). Isolated endotoxins have always been reported to contain various amounts of proteins, as well as of lipopolysaccharides. The presence of extractable phospholipids has not always been investigated, and where they are reported they amount to between 10 and 28 per cent of the endotoxins, the highest reported proportions being found consistently in the excreted endotoxins of *E. coli* 12408 (Knox, Cullen and Work, 1967; Rothfield and Pearlman-Kothencz, 1969). It is therefore possible that proteins, phospholipids and lipopolysaccharides may either exist together in the cell wall as a complex, or are lost from the wall simultaneously and then form complexes. Walls of gram-negative bacteria contain about 30–40 per cent of their weight as lipopolysaccharides and phospholipids, and a close association between them *in vivo* is expected in view of the demonstration by Rothfield and Takeshita (1966) that *in vitro* the biosynthetic addition of sugars to incomplete lipopolysaccharides only occurs when the lipopolysaccharide has been previously "annealed" with a phospholipid such as phosphatidylethanolamine.

The excreted endotoxin originates from the outer layers of the cell walls. In lysine-limited *E. coli* production of excreted endotoxin was not accompanied by cell lysis, but apparently arose through over-growth of the outer layers of the walls, thus forming blebs which were then pinched off and lost into the surrounding medium (Work, Knox and Vesk, 1966; Knox, Vesk and Work, 1966; Work, 1967). Such cultures showed a biphasic growth; when lipopolysaccharide was specifically estimated by immunological methods, extracellular production was found to commence on depletion of lysine from the culture medium and to continue at a constant rate right through stationary and secondary phases of growth. The actual mechanism of production is not known; Rothfield and Pearlman-Kothencz (1969) confirmed this excretion of the lipopolysaccharide-containing complex: they suggested that it was not a specific consequence of lysine limitation but a more general result of cessation of protein synthesis, since about 50 per cent of the excretion also occurred in cultures grown in the presence of chloramphenicol. However, growing cells of *E. coli* 12408 given adequate lysine also excreted very small amounts of lipopolysaccharide, so one cannot be sure that one is dealing with the same phenomenon in all cases.

The use of ferritin-labelled antibodies indicated that endotoxins are in the outer layers of the walls (Shands, 1965; Knox, Vesk and Work, 1966), although there is still doubt as to whether protein or lipopolysaccharide singly make up the outermost layer or whether they are present as a network.

The removal of endotoxin from cells by such agents as salts or ether (methods 4 and 5, Table I) suggests that the linkages binding the endotoxin components to the rest of the wall are not covalent.

Lipopolysaccharides show a great tendency to form aggregates or to complex with other molecules, and it is probable that even in the cell wall there are lipopolysaccharide-protein complexes and that the complexes are not artifacts produced during excretion or extraction. Calcium or other divalent cations may also be involved in cross-linking lipopolysaccharide subunits and in binding them to other components of the wall (Asbell and Eagon, 1966). This would account for the removal of endotoxins from bacterial cells by EDTA (Leive, 1965) which could strip the cations from the walls. Divalent cations are known to complex easily with lipopolysaccharides and endotoxins *in vitro* to form insoluble precipitates (Osborn *et al.*, 1962; Knox, Cullen and Work, 1967).

Some, if not all, of the lipopolysaccharide may be linked indirectly to the mucopeptide component of cell walls. In *E. coli* a lipoprotein has been found bound to diaminopimelic acid residues of the mucopeptide through lysine (Braun and Rehn, 1969). It is not known whether this lipoprotein occurs in the endotoxin complex; if it does, it might account for the "excretion" of endotoxin by lysine-limited cells. It has been reported (Raynaud *et al.*, 1966; Raynaud, Chermann and Digeon, 1969) that cells treated with sodium dodecylsulphate during and after harvesting will not yield up their endotoxin or lipopolysaccharide by the usual extraction procedures. Sodium dodecylsulphate is known to inhibit muralytic enzymes, and it was suggested that the preliminary degradation of mucopeptides by these enzymes which can occur during harvesting of cells is necessary for successful extraction of endotoxins or lipopolysaccharides.

COMPOSITION OF ENDOTOXIN COMPONENTS

Since lipopolysaccharides are known to be the most active component of endotoxins, more attention has been devoted to their composition than to that of the protein and phospholipid components. Various experiments have indicated that the three components exist as a "complex" which however does not make up all of the crude endotoxin preparations (Janda and Work, 1970; Rothfield and Pearlman-Kothencz, 1969; Rogers, Gilleland and Eagon, 1969). The criteria used for the existence of a complex rather than a mixture are: occurrence of all three components in at least one of the fractions obtained when the endotoxins are subjected to various physical methods of separation such as density-gradient centrifugation, gel electrophoresis or gel filtration. No single one of these is really indicative of more than a close physical association of the components, but is more meaningful when taken in conjunction with the others.

The main phospholipid (phosphatidylethanolamine) has been purified from excreted endotoxin of lysine-limited *E. coli* 12408 and has been fully characterized (Knox, Cullen and Work, 1967; Work, 1969; van Golde and van Deenen, 1967). It is noticeable that the main fatty acids (C_{16}, C_{18}) are not the same as the "bound" fatty acids of lipid A (C_{12}, C_{14}), pointing to a different origin of the two types of lipids. However we detected no gross difference in phospholipid fatty acids of endotoxins and complete cell walls. Phospholipids of many other genera are also mainly phosphatidylethanolamine, but their fatty acid composition may be different from that of *E. coli*.

The proteins of endotoxins are probably not homogeneous and are insoluble when separated from the complexes, therefore results of chemical analysis of them are less meaningful. However a few specimens examined had, in common with many extracellular proteins, an excess of acidic over basic amino acids and little if any cystine (Homma and Suzuki, 1964; Knox, Cullen and Work, 1967; Crutchley, Marsh and Cameron, 1968). While in the endotoxin complex, protein is not digestible by pronase (Rothfield and Pearlman-Rothencz, 1969; Raynaud, Chermann and Digeon, 1969), but removal of lipopolysaccharide and phospholipids renders it pronase-sensitive.

Separated lipopolysaccharides are also inhomogeneous (e.g. Nowotny, 1966). Nevertheless, it has been possible to construct an overall picture of the lipopolysaccharide molecule of various organisms, particularly of the salmonellas (for reviews see Lüderitz, Staub and Westphal, 1966; Lüderitz, Jann and Wheat, 1968; Lüderitz, 1970). About 30 different sugar constituents have been identified in various lipopolysaccharides. Most types of lipopolysaccharides contain at least five sugars, and often more; glucose, galactose, glucosamine, L-glycero-D-mannoheptose, and 2-keto-3-deoxyoctonic acid (KDO) are most common and together with phosphate groups and ethanolamine constitute the "core polysaccharide" found in most Enterobacteriaceae. The core is one of three regions of the complete lipopolysaccharide unit (Fig. 1); another region consists of

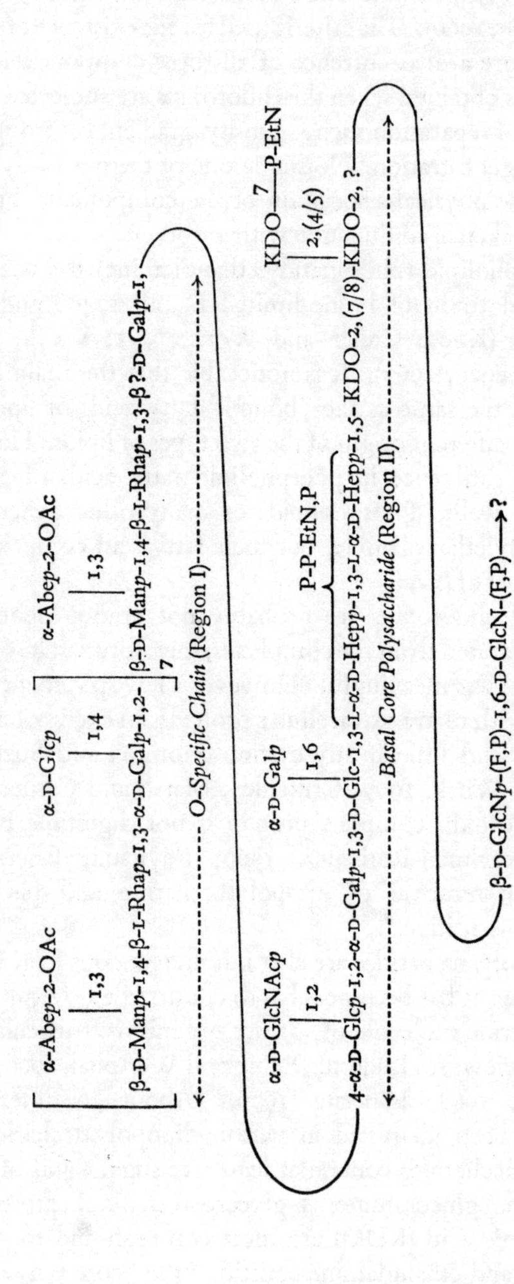

FIG. 1. Composition of a lipopolysaccharide from *Salmonella typhimurium* (Lüderitz, 1970).

O-specific side chains carrying the immunological specificities of the O-antigens; the third region is lipid A. The O-side chains constitute repeating units of oligosaccharides which contain specific sugars in specific linkages, thus differentiating the lipopolysaccharides of the various serotypes. They are characteristic of smooth (S) type pathogens, and are absent from laboratory mutant strains of organisms, known as rough (R) strains. (The terminology is based on the appearance of colonies on agar plates.) With the exception of O-antigenicity, rough lipopolysaccharides retain all the biological properties of the complete O-antigenic polymers, including endotoxic and pyrogenic properties.

The structure of the core polysaccharide is probably identical, or very similar, in the various salmonella serotypes; but in other genera such as *Shigella* or *Escherichia* there are some structural differences, in that certain sugars may be missing (e.g. galactose or glucosamine). A fairly common part of each core polysaccharide of Enterobacteriaceae contains two molecules of heptose linked on one side (the left in Fig. 1) to the oligosaccharide containing one or more residues of glucose, galactose and glucosamine; the other side is linked to a branched trisaccharide of KDO. There are also two phosphate groups in the heptose portion, but it is not known whether or not they act as bridges to neighbouring chains. Ethanolamine or phosphoethanolamine is also found in both the heptose and the KDO portions of the core. KDO joins the core to lipid A by a link which is very acid-labile but relatively stable to alkalis, suggesting that it is a ketosidic link to the glucosamine in the lipid A.

Weak acid hydrolysis splits the lipopolysaccharide molecule into soluble degraded polysaccharide, free KDO plus ethanolamine phosphate, and insoluble lipid A. None of these fractions when pure have endotoxic activity, although the degraded polysaccharide is haptenic. The composition and general structure of lipid A preparations from various bacterial genera seem to be closely related, although there are individual differences in fatty acid content. Crude lipid A obtained by hydrolysis of lipopolysaccharide represents a mixture of fractions which can be separated by various means (Nowotny, 1963*a*; Kasai, 1966; Burton and Carter, 1964). A possible structure for lipid A of *S. minnesota* and *Serratia marcescens* has been suggested by Gmeiner, Lüderitz and Westphal (1969) and Adams and Singh (1970) (see Fig. 2). It consists of several glucosamine residues bound by β,1→6 linkages, with most of the hydroxyl groups substituted by fatty acids (C_{12}, C_{14}, C_{16}) or acetyl groups, and with the amino groups substituted by β-hydroxymyristic acid. The specific and characteristic component of these and other lipid A preparations is β-hydroxymyristic acid; in some genera the lower homologue, β-hydroxylauric acid, may

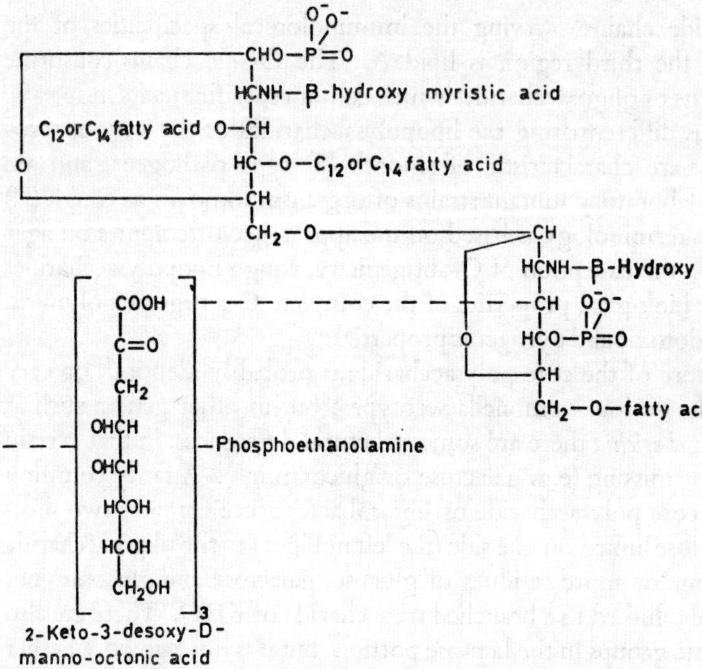

FIG. 2. Structure of the "glycolipid" of a *Salmonella* Re mutant.

be present in addition or exclusively (Fensom and Gray, 1969; Hancock, Humphreys and Meadow, 1970).

Various R mutants of altered serological specificity are known to exist, some naturally occurring, others produced artificially. Such mutants are known to be defective in one of the gene loci on the chromosome which determines distinct steps of lipopolysaccharide biosynthesis. These mutant cells therefore synthesize defective lipopolysaccharide but are quite viable under laboratory conditions. There are two main groups of R mutants; those with defective synthesis of O-specific side chains (known as Ra mutants) which produce complete lipopolysaccharide except for the O side chain, and those defective in some step in the synthesis of the basal core (Rb-Re mutants). Table III shows diagrammatically the scheme of

TABLE III

LIPOPOLYSACCHARIDES FROM ROUGH MUTANTS OF *Salmonella minnesota*

Chemotype	*Glucosamine*	*Galactose*	*Glucose*	*Heptose*	*KDO*	*Lipid A*
Ra	+	+	+	+	+	+
Rb		+	+	+	+	+
Rc			+	+	+	+
Rd				+	+	+
Re*					+	+

*Glycolipid

classification and composition of a series of lipopolysaccharides from *S. minnesota* R mutants which have helped considerably to establish lipopolysaccharide composition. Ra type lipopolysaccharides represent the complete basal core; Rb to Re mutants produce progressively defective cores. Lipopolysaccharides from Re mutants which are devoid of hexoses and heptoses (heptose-less), and are composed only of KDO, ethanolamine, phosphorus and lipid A, (Fig. 2) are referred to as "glycolipids". They retain the full endotoxic and pyrogenic activities of the complete lipopolysaccharide molecule; but, being very lipophilic, they are difficult to dissolve in water or to purify (Lüderitz *et al.*, 1966; Kasai and Nowotny, 1967; Nowotny, Kasai and Tripodi, 1969; Kim and Watson, 1967).

The pyrogenic activities of endotoxins have been shown to be independent of the presence not only of proteins and phospholipids but also of O-specific chains and the core polysaccharide—except for the KDO portion. When KDO is removed from the glycolipid by mild acid hydrolysis the free lipid A is said to be inactive. Thus, although lipid A is essential for pyrogenic and other activities, some or all of the additional polar groups of glycolipid such as KDO and phosphatidylethanolamine must be necessary for full activity. Since esterification of all free hydroxyl groups in the glycolipid did not lead to changes in endotoxic properties (McIntire *et al.*, 1967) these groups may not be essential for toxicity, although they may be concerned in serological reactivity as KDO is the immunodominant sugar in both glycolipid (Risse *et al.*, 1967) and excreted lipopolysaccharide of *E. coli* 12408 (Knox, 1966).

PROPERTIES OF ENDOTOXINS AND LIPOPOLYSACCHARIDES

Physical properties

Endotoxins, whether extracted or excreted, are very large molecules with a molecular weight of the order of 1–2 millions. They are water-soluble if they have not been previously lyophilized. Lipopolysaccharides may be aggregates of small subunits and have molecular weights of 800 000–1 000 000, and they are easily sedimentable at high speeds. S-type lipopolysaccharides are usually soluble in water, but R-type molecules are more difficult to dissolve especially after lyophilization. Solutions of both endotoxins and lipopolysaccharides are strongly opalescent; the opalescence disappears after treatment with surface-active agents such as sodium dodecylsulphate or with bases such as triethylamine or dilute sodium hydroxide. It is customary to solubilize lyophilized preparations with the aid of dilute alkalis, but the wisdom of this is questionable (see

p. 39). For the preparation of homogeneous, completely unaltered endotoxins, only the mildest possible methods of isolation should be used and lyophilization should be avoided.

Biological properties

The diverse nature of the characteristic biological properties of endotoxins is summarized by Nowotny (1969). Death is, of course, a non-specific response to many different types of toxins, while the development of fever is to some extent a specific physiological response to endotoxins. It is generally accepted that the pyrogenic potency of endotoxins reflects their numerous other toxic properties, but that serological reactivity, and stimulation of the host defence mechanisms (immunogenicity, adjuvant effect, and promotion of non-specific resistance) are unrelated to their other biological properties.

The typical biphasic fever curve with peaks at approximately one and three hours produced by suitable intravenous doses of endotoxins is considered to be the nearest to a biological response definitive for endotoxins (Milner and Finkelstein, 1966). However, the elaborate set-up for accurate pyrogen testing is not always available to all workers in the field. A cheap, reasonable alternative test is that of chick embryo toxicity, which was shown by Milner and Finkelstein (1966) to be interchangeable with rabbit pyrogenicity tests for endotoxins, but which does not give identical results on detoxified preparations (Cundy and Nowotny, 1968).

There is no doubt that molecular size, solubility, and degree of dispersion of test preparations can influence biological responses, particularly pyrogenicity. Polymers or particles unrelated to typical endotoxins may show pyrogenic activities, e.g. dextrans, glycogen, inorganic colloids, viruses, fat emulsions, or gram-positive bacterial cells (Atkins, 1960). Endotoxins showed eight times higher pyrogenic activity when extracted from cell walls than when the whole walls were tested (Milner and Finkelstein, 1966), which suggests that solubilization had increased pyrogenicity. The glycolipid from heptose-less Re mutants of *S. minnesota* is not soluble in water, and its pyrogenicity was enhanced tenfold by dispersing it in water with triethylamine (Westphal *et al.*, 1969). In these last two examples, increases in pyrogenicity over the original preparations were higher than increases in toxicity.

Another biological property of endotoxins and lipopolysaccharides is the ability to produce tolerant states when administered in repeated sublethal doses. This tolerance is non-specific; endotoxins of one serological type induce cross-tolerance to serologically unrelated endotoxins. These properties differentiate gram-negative endotoxins from some non-endotoxic pyrogens; some pyrogens from pathogenic fungi do not induce any tolerance (Braude, McConnell and Douglas, 1960), while with others tolerance is very specific (Stetson, 1956). Endotoxin tolerance is particularly noticeable in pyrogenic responses in animals and humans (Favorite and Morgan, 1942; Beeson, 1947; Watson and Kim, 1963; Kim and Watson, 1967). Progressive decreases in height and duration of the febrile responses result from frequent injections of endotoxins, and the tolerant fever curve finally becomes monophasic, showing no secondary response at three hours.

The part of the molecule inducing tolerance is evidently the same as that responsible for endotoxicity, since animals rendered tolerant to glycolipid, which is devoid of all O and R polysaccharides, were also immune to the pyrogenic effects of the complete lipopolysaccharides derived from wild-type parent cells, or to heterologous endotoxins from smooth or rough organisms (Kim and Watson, 1967).

HETEROGENEITY

The classical concepts (if they now exist) of absolute physical and chemical homogeneity cannot be applied strictly to such unusually large and complex polymers as endotoxins or lipopolysaccharides, and we must accept that trace contaminants and minor structural changes or components may not be revealed by any analytical methods now available. Biological homogeneity is also difficult to ensure, or inhomogeneity to eliminate, owing to the imprecision of our methods of testing. However, with the methods available today, it is certain that there is great physical, chemical and biological heterogeneity in all preparations used in pyrogenic or other tests—whether they are original endotoxins, purified lipopolysaccharides or glycolipids. The conflicting reports, now largely resolved, as to the endotoxic nature of the lipid A moiety of lipopolysaccharides can be attributed to inhomogeneity of the test preparations. Lipid A is not a compound, it is just the water-insoluble *fraction* obtained by mild acid hydrolysis of lipopolysaccharide (Westphal and Lüderitz, 1954) and has been shown by Nowotny (1963*a*) and Kasai (1966) to contain at least 16 components. Some 5–10 per cent of the original endotoxic or pyrogenic activity was retained by unfractionated lipid A preparations (Westphal *et al.*, 1958) and the activity was slightly enhanced when the lipid was solubilized by heating with casein. Ribi and co-workers (1961) did not agree as to the essential nature of lipids in endotoxic products. With our knowledge of the chemistry of glycolipids obtained with the help of mutants (see p. 33), and with chemical analyses of lipid A fractions (Nowotny, 1969; Adams and Singh, 1970), it is now possible to attribute residual activity to incompletely degraded lipopolysaccharide fragments containing KDO and other core constituents. This is to be expected, since lipid A is obtained by partial acid hydrolysis which would yield a mixture of barely altered polysaccharide core material, glucosamine-linked fatty acids and all intermediate fragments; some of these will be found in the water-insoluble material known as lipid A.

Lipopolysaccharide preparations are themselves known to be heterogeneous, as shown by ion-exchange chromatography and density-gradient centrifugation (Nowotny, 1966; Nowotny *et al.*, 1966). It is uncertain whether the comparatively rigorous hot aqueous phenol method used for separation of lipopolysaccharides causes heterogeneity, or whether many species of lipopolysaccharides exist in the wall. Certainly, chemical heterogeneity is now becoming apparent in lipopolysaccharides both in sugar contents (Holme *et al.*, 1969) and in the KDO moiety (Janda and Work, 1970). We examined alkali-treated excreted endotoxin of *E. coli* 12408 and its separated lipopolysaccharide for reactivity in the periodate thiobarbituric acid colour test as used in the KDO estimation, and found that the proportion of the KDO reacting directly (without preliminary acid hydrolysis) was lower in material excreted during the first stationary phase than in the materials produced during secondary growth. We have also found (unpublished) heterogeneity in the lipopolysaccharides of fractions obtained by gel filtration of excreted endotoxin.

Density-gradient centrifugation has separated endotoxins into fractions of differing morphology, and differing chemical, serological and endotoxic properties (Beer *et al.*, 1965; Beer, Braude and Brinton, 1966; Nowotny *et al.*, 1966; Work, 1969). Some of our results on excreted endotoxin (Table IV) resemble those reported on extracted endotoxins, and show that there was considerable variation in protein and lipopolysaccharide contents of materials sedimenting through the gradient, and also in their chick embryo toxicities and serological activities. Unfractionated "Boivin type" endotoxins from *Serratia marcescens* contain a mixture of nucleic acid, acidic polysaccharides and endotoxin-like complexes of varying sizes (Alaupovic, Olson and Tsang, 1966). The acidic polysaccharides showed some pyrogenic activity, but some contamination with lipopolysaccharide was not ruled out.

Several protein fractions have been separated from the extracellular culture filtrate of *E. coli* 12408 (Janda and Work, 1970); some of these are excreted under all conditions while others are more specifically associated with intense lipopolysaccharide excretion. In general, both extracted and excreted types of endotoxins are likely to contain more than one type of

TABLE IV

DENSITY GRADIENT FRACTIONATION OF EXCRETED ENDOTOXIN FROM *E. coli* 12408†

Fraction	Protein (%)	Phosphorus (%)	Heptose (%)	Egg toxicity LD_{50} (μg)	Reaction in gels with antiserum to: Endotoxin	Protein	Whole cells
1	34	1·5	7		+	+	+
3	25	2·5	12		+	+	+
4	16	4·0	16	16	±	0	+
5	12	4·0	16	5	0*	0	+
6	8	3·0	16	2	0*	0	+
7 (pellet)	9	4·0	17	9	0*	0	+
Unfractionated endotoxin	10	3·2	10	8	+	+	+

* Positive after pretreatment with sodium dodecylsulphate (0·05% w/v).
† From Work (1969).

protein. Some endotoxin proteins have been reported to have toxic properties (Homma and Suzuki, 1964; Raynaud, Chermann and Digeon, 1969; Goebel and Barry, 1958). It is still uncertain whether these toxicities are due to residual undetected contamination with lipopolysaccharide or glycolipid or if they are inherent in the protein molecule itself. Thus, acid treatment of endotoxin extracted by diethylene glycol produced a toxic protein fraction (Binkley, Goebel and Perlman, 1945) which probably contains lipid A and/or degraded lipopolysaccharide complexed with protein. This certainly suggests that the lipopolysaccharide is attached to protein in the endotoxin complex through the KDO-lipid A moiety. Whether the links are covalent or electrostatic is not known. Certain KDO molecules (possibly those in the side chain of the trisaccharide unit) may lower the capacity for aggregation or complex formation shown by the rest of the lipopolysaccharide polymer. Thus, mild acid treatment (pH 4·6) leading to aggregation resulted in the release of free KDO from the excreted lipopolysaccharide of *E. coli* 12408 (Knox, 1966; Work, 1969). For example, after 30 minutes at 100°C (pH 4·6) the sedimentation coefficient had increased from 4·1S to 9·5S and on standing overnight a gel settled out. The precipitation of lipid A only occurred after more prolonged heating when gel formation was decreased, presumably due to degradation of the polymer. The removal of 80 per cent of this easily-released KDO residue by heating for only 10 minutes resulted in loss of serological activity but only a slight drop in chick embryo toxicity: precipitability by magnesium was lost, suggesting that this KDO group may be responsible for divalent cation binding or for the molecular configuration causing binding. Alternatively this loss of affinity for magnesium could have been the consequence of the aggregation covering the magnesium binding groups. It is probable that treatment of lipopolysaccharides under unsuitable conditions (possibly even when complexed with proteins) may result in intramolecular changes through breakage of labile linkages. These changes, by causing aggregation or disaggregation, can produce an inhomogeneous population of polymer molecules.

DETOXIFICATION OF ENDOTOXINS

Since endotoxins and lipopolysaccharides are usually obtained as turbid solutions of aggregates of extremely high molecular weight, one should consider whether they are active *in vivo* as such, or whether they are disaggregated in the blood stream. The fact that endotoxins and lipopolysaccharides can be reversibly disaggregated into subunits of much reduced molecular weight by surface-active agents enables this aspect to be investigated. Sodium deoxycholate disaggregates Boivin endotoxin or extracted lipopolysaccharide into non-pyrogenic subunits; on removal of deoxycholate by dilution or dialysis, reaggregation occurs and pyrogenicity is restored (Ribi *et al.*, 1966; McIntire *et al.*, 1969). The presence of serum or other proteins inhibits restoration of pyrogenicity and reaggregation. The results of McIntire and co-workers indicate that loss of pyrogenicity was not due to the disaggregation, since the same degree of disaggregation into the subunits of the same size was produced both by low concentrations of deoxycholate, which did not affect pyrogenicity, and by higher concentrations, which abolished pyrogenicity completely. The effect of deoxycholate on pyrogenicity must therefore have been caused by reversible blocking of the pyrogenic portion of the molecule or by changing its conformation, not by disaggregating the lipopolysaccharide polymer. This is supported by the fact that sodium dodecylsulphate strongly dissociates lipopolysaccharides but does not eliminate pyrogenicity (Beer, Braude and Brinton, 1966; Nowotny, 1969; McIntire *et al.*, 1967).

Lack of connexion between particle size and pyrogenicity or toxicity is confirmed by the results of mild alkaline treatment, which is used to solubilize lipopolysaccharides and also to render them capable of coating erythrocytes for subsequent haemagglutination tests. Treatment of endotoxins at room temperature with 0·1N-NaOH produced an immediate drop in particle size (estimated by light scattering) from 9×10^6 to 3×10^6; less rapid changes (starting at three hours) in molecular dissymmetry and toxicity were found and after eight hours no toxicity was apparent (Tripodi and Nowotny, 1966). Other workers have also found that treatment with alkali eliminates endotoxicity or pyrogenicity (McIntire *et al.*, 1967; Marx *et al.*, 1968; Neter *et al.*, 1956). Liberation of free fatty acids (other than β-hydroxymyristic acid in some cases) occurred during alkali treatment and additional hydroxyl groups appeared. Unless other undetected chemical changes occurred, this suggests that removal of some of the ester-linked fatty acids had eliminated endotoxicity. Tripodi and Nowotny (1966) suggest that this then produced unfolding or swelling of the molecule (revealed by increase in dissymmetry), leading to distortion of the original toxic structure and consequent loss of activity. It now seems important to study changes on individual residues of the lipopolysaccharide molecule. For example, the appearance of a periodate-thiobarbiturate-reactive KDO group after excreted endotoxin or lipopolysaccharide had been in contact with 1·0N-NaOH for 30 minutes at room temperature (Janda and Work, 1970) suggests either that a hydroxyl group on a KDO residue has been unmasked through a deacylation reaction, or that disaggregation has increased the availability of these hydroxyl groups for reaction. Whatever the mechanism of mild alkaline detoxification, it is impressive in its rapidity; it is also interesting that the losses in different biological activities did not occur simultaneously (Cundy and Nowotny, 1968).

Much effort has been put into finding other methods of reducing the pyrogenicity and toxicity of endotoxins without removing their more desirable qualities which make them suitable for vaccines. Acetylation reduced pyrogenicity at least 100-fold (Keiderling, Wöhler and Westphal, 1953; Martin and Marcus, 1966; Freedman and Sultzer, 1962); on deacetylation activity was recovered by Freedman and Sultzer. Succinylation had little effect on pyrogenicity (McIntire *et al.*, 1967). The following treatments produced products (termed "endotoxoids" by Nowotny, 1963*b*) with reduced pyrogenicity or toxicity: lithium aluminium hydride (Noll and Braude, 1961), transesterification with boron trifluoride, O-acyl cleavage with potassium methylate, treatment with pyridine and formic acid (Nowotny, 1963*b*; Johnson and Nowotny, 1964). All these procedures were found to split ester-bound carboxylic acids from model compounds, and the fatty acid contents of the "endotoxoid" preparations were reduced.

Some detoxified endotoxins, although themselves non-pyrogenic, are still capable of inducing pyrogenic tolerance. Animals injected with non-pyrogenic lithium aluminium hydride-treated endotoxin do not develop fever when subsequently injected with untreated endotoxin (Noll and Braude, 1961). On the other hand, non-pyrogenic immunologically-active acetylated endotoxins or lipopolysaccharides do not induce tolerance (Freedman and Sultzer, 1962; Keiderling, Wöhler and Westphal, 1953).

The action of serum in lowering pyrogenicity of endotoxins (e.g. Hegemann and Lessmann, 1958; Lüderitz *et al.*, 1958) may be due at least partly to enzymic action (Keene *et al.*, 1961), although reversible complexing by various serum proteins may occur (Rudbach and Johnson, 1966). Two serum fractions were implicated in enzymic detoxification by Skarnes (1966, 1968); one of these (α_1-lipoprotein esterase) degraded endotoxin but did not inactivate it, while the other (α_1-globulin esterase) destroyed toxicity but not serological activity. The rate of detoxification was inversely proportional to the serum concentration of calcium ions and it is suggested that the endotoxin itself may be binding calcium, so causing a sufficient drop in serum Ca^{++} to activate its own destruction. A completely different type of biological degradation was achieved by Malchow and co-workers (1969) who found that amoebae of the cellular slime mould *Dictyostelium discoideum* when grown in submerged culture on gram-negative bacteria as the sole nutrient released into the medium a partially-degraded lipopolysaccharide of mol. wt.

15 400. This substance contained all the sugars of the parent lipopolysaccharide including the glucosamine of lipid A, but had no long-chain fatty acids. The degraded polysaccharide retained its serological specificity, but failed to sensitize erythrocytes for passive haemagglutination, and had LD_{50} = 100 μg per mouse compared with LD_{50} = 2 μg for the untreated lipopolysaccharide. Unfortunately no figures were published for pyrogenicity.

It seems possible that these biological degradations may be producing the same type of molecules as the more unspecific chemical methods of detoxification.

HEAT STABILITY OF LIPOPOLYSACCHARIDES

Although generally regarded as heat-stable compounds, some changes do occur in heated lipopolysaccharides. Thus Neter and co-workers (1956) found that heating for 2½ hours at 100°C at pH 7·3 produced noticeable changes in the ability to coat erythrocytes. I have found (Work, unpublished) that the toxicity for chick embryos of an autoclaved solution (1 mg/ml) of excreted lipopolysaccharide from *E. coli* 12408 changed on storage at 2°C (Fig. 3). Toxicity was lost steadily for six weeks, after which

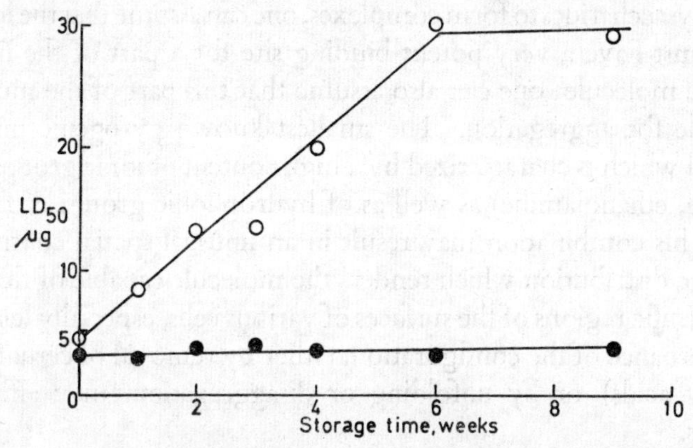

FIG. 3. Chick embryo toxicity of stored lipopolysaccharide (1 mg/ml) from excreted endotoxin of *E. coli* 12408. ○: autoclaved; ●: not autoclaved (Work, unpublished).

there was no further change; control unautoclaved material (sterilized by filtration) was unaffected. This experiment was carried out to investigate the reason for losses in pyrogenicity of our autoclaved lipopolysaccharide which P. A. Murphy (private communication) had found in successive experiments, and it emphasizes that caution should be exercised in preparing standards for pyrogenic testing. The slowness of the inactivation is remarkable, and suggests that the alteration of the lipopolysaccharide by autoclaving may have rendered it more susceptible either to absorption by the glass surface of the bottle, or to degradation by alkali leached out of the glass on prolonged standing.

MOLECULAR BASIS FOR BIOLOGICAL ACTION OF ENDOTOXINS

I have reviewed briefly the work of many investigators over many years, and I think that the quotation before my opening paragraph is still as true today as it was when spoken by Dr I. L. Bennett when opening the Symposium on Endotoxins at Rutgers University in 1964. In spite of all the new facts that have accumulated since then, the molecular basis for biological action of endotoxins still "just eludes detection"; endotoxins still have a fabulous intrinsic fascination and are still irresistible—certainly to me and my colleagues.

The pyrogenic action of lipopolysaccharides is probably partly a secondary effect of their reaction with the leucocyte, a theory which is at least quantitatively satisfying, as Westphal and co-workers (1955) showed; the pyrogenic dose for man (0·1 μg) is about 10^{11} molecules, which is approximately the number of circulating leucocytes. In view of the great tendency of lipopolysaccharides to form complexes, one can assume that the leucocyte surface must have a very potent binding site for a part of the lipopolysaccharide molecule; one can also assume that this part of the molecule is responsible for aggregation. The smallest known pyrogenic unit is the glycolipid which is characterized by a high content of ionic groups (KDO, phosphate, ethanolamine) as well as of hydrophobic groups of fatty acid chains. This combination may result in an unusual spatial configuration and charge distribution which renders the molecule capable of fitting into certain specific regions of the surfaces of various cells, especially leucocytes. Any disturbance of the configuration, either by removal of certain groups (e.g. fatty acids) or by unfolding or disaggregation, may affect toxic properties.

OTHER PYROGENS

Other natural pyrogens differ from the endotoxins in many respects and little work has been carried out on their chemistry. Purified preparations of toxin from *Streptococcus pyogenes* are pyrogenic in doses of about 10 μg per rabbit; they give typical biphasic fever curves and induce tolerance, but the tolerance is very specific and thus differs from that induced by endotoxins (Watson, 1960; Stetson, 1956). The pyrogenic effects of the streptococcal toxin can be neutralized by specific antiserum, unlike that of endotoxins (Kim and Watson, 1966).

Pyrogenic responses occur to very high doses of cells or filtrates of pathogenic fungi such as *Candida albicans*, *Histoplasma capsulatum* and *Blastomyces dermatitidis* (Braude, McConnell and Douglas, 1960). A dose of 10^6 cells of *C. albicans* only produced a temperature rise of 0·5–0·8°F (0·28–0·44°C) and there was no induction of tolerance. An "endotoxic-like" fraction from *C. albicans* was immunogenic and lethal but was not reported to be pyrogenic or to contain lipid (Isenberg, Allerhand and Berkman, 1963).

Polysaccharides prepared from such diverse sources as mice tissues, chick embryos, tangerines and *Bryonia* roots showed varying degrees of pyrogenicity (minimum pyrogenic dose varying from 0·2 to 50 μg) and had other endotoxic properties, but they induced a non-specific tolerance (Landy and Shear, 1957). Although this work does not point to the presence of true "endotoxins" in tissues, it does suggest that caution should be exercised when attempting to measure or identify bacterial endotoxins in biological materials.

The pyrogenic effects of other polymers have already been mentioned. However, recently, full endotoxic properties have been found in the double-stranded semi-synthetic RNA polymer, polyinosinic-polycytidylic acid (poly I·poly C) (Weinstein, Waitz and Came, 1970; Absher and Stinebring, 1969). Poly I·poly C in doses of 0·5 μg/kg gave a typical biphasic temperature rise of 0·8°C; the single-stranded constituents, poly I or poly C had no effects at levels of 25 μg/kg. Non-specific tolerance was induced, poly I·poly C producing tolerance to its own toxic effects or to those of gram-negative bacterial endotoxins, while endotoxins gave tolerance to poly I·poly C. No data are available at present as to the mechanisms of these effects, but since they are produced by substances of known composition, it is possible that further work may help in elucidating the mode of action of endotoxic pyrogens.

REFERENCES

ABSHER, M., and STINEBRING, W. R. (1969). *Nature, Lond.*, **223**, 715–717.
ADAMS, G. A., and SINGH, P. P. (1970). *Can. J. Biochem.*, **48**, 55–63.
ALAUPOVIC, P., OLSON, A. C., and TSANG, J. (1966). *Ann. N.Y. Acad. Sci.*, **133**, 546–565.
ASBELL, M. A., and EAGON, R. G. (1966). *J. Bact.*, **92**, 380–387.
ATKINS, E. A. (1960). *Physiol. Rev.*, **40**, 580–646.
BEER, H., BRAUDE, A. I., and BRINTON, C. C. (1966). *Ann. N.Y. Acad. Sci.*, **133**, 450–475.

BEER, H., STAEHELIN, T., DOUGLAS, H., and BRAUDE, A. I. (1965). *J. clin. Invest.*, **44**, 592–602.
BEESON, P. B. (1947). *J. exp. Med.*, **86**, 29–44.
BENNETT, I. L. (1964). *Bacterial Endotoxins*, p. xiii, ed. Landy M., and Braun, W. New Brunswick: Rutgers University Press.
BINKLEY, F., GOEBEL, W. F., and PERLMAN, E. (1945). *J. exp. Med.*, **81**, 331–347.
BISHOP, D. G., and WORK, E. (1965). *Biochem. J.*, **96**, 567–576.
BOIVIN, A., and MESROBEANU, L. (1933). *C.r. Séanc. Soc. Biol.*, **114**, 307–310.
BRAUDE, A. L., MCCONNELL, J., and DOUGLAS, H. (1960). *J. clin. Invest.*, **39**, 1266–1276.
BRAUN, V., and REHN, K. (1969). *Eur. J. Biochem.*, **10**, 426–438.
BURTON, A. J., and CARTER, H. E. (1964). *Biochemistry, Wash.*, **3**, 411–418.
CORPE, W. A., and SALTON, M. R. J. (1966). *Biochim. biophys. Acta*, **124**, 125–135.
COX, S. T., and EAGON, R. G. (1968). *Can. J. Microbiol.*, **14**, 913–922.
CRUTCHLEY, M. J., MARSH, D. G., and CAMERON, J. (1968). *J. gen. Microbiol.*, **47**, 413–420.
CUNDY, K. R., and NOWOTNY, A. (1968). *Proc. Soc. exp. Biol. Med.*, **127**, 999–1003.
FAVORITE, G. O., and MORGAN, H. J. (1942). *J. clin. Invest.*, **21**, 589–599.
FENSOM, A. H., and GRAY, G. W. (1969). *Biochem. J.*, **114**, 185–196.
FREEDMAN, H. H., and SULTZER, B. M. (1962). *J. exp. Med.*, **116**, 929–942.
GALANOS, C., LÜDERITZ, O., and WESTPHAL, O. (1969). *Eur. J. Biochem.*, **9**, 245.
GMEINER, J., LÜDERITZ, O., and WESTPHAL, O. (1969). *Eur. J. Biochem.*, **7**, 370–379.
GOEBEL, W. F., and BARRY, G. T. (1958). *J. exp. Med.*, **107**, 185–209.
GOEBEL, W. F., BINKLEY, F., and PERLMAN, E. (1945). *J. exp. Med.*, **81**, 315–330.
GOLDE, L. M. G. VAN, and DEENEN, L. L. M. VAN (1967). *Chem. Phys. Lipids*, **1**, 157–164.
HANCOCK, I. C., HUMPHREYS, G. O., and MEADOW, P. M. (1970). *Biochim. biophys. Acta*, **202**, 389–391.
HASKINS, W. T., LANDY, M., MILNER, K. C., and RIBI, E. (1961). *J. exp. Med.*, **114**, 665–684.
HEGEMANN, F., and LESSMANN, H. (1958). *Z. ImmunForsch. exp. Ther.*, **115**, 391–401.
HOLME, T., LINDBERG, A. A., GAREGG, P. J., and ONN, T. (1969). *Colloques int. Cent. natn. Rech. Scient.*, **174**, 121–131.
HOMMA, J. V., and SUZUKI, N. (1964). *J. Bact.*, **87**, 630–640.
ISENBERG, H. D., ALLERHAND, J., and BERKMAN, J. I. (1963). *Nature, Lond.*, **197**, 516–517.
JANDA, J., and WORK, E. (1970). *J. gen. Microbiol.*, **61**, vii.
JOHNSON, A. G., and NOWOTNY, A. (1964). *J. Bact.*, **87**, 809–814.
KASAI, N. (1966). *Ann. N.Y. Acad. Sci.*, **133**, 486–507.
KASAI, N., and NOWOTNY, A. (1967). *J. Bact.*, **94**, 1824–1836.
KEENE, W. R., LANDY, M., SHEAR, M. J., and STELECKY, K. A. (1961). *J. clin. Invest.*, **40**, 302–310.
KEIDERLING, W., WÖHLER, F., and WESTPHAL, O. (1953). *Arch. exp. Path. Pharmak.*, **217**, 293–311.
KIM, Y. B., and WATSON, D. W. (1966). *Ann. N.Y. Acad. Sci.*, **133**, 727–745.
KIM, Y. B., and WATSON, D. W. (1967). *J. Bact.*, **94**, 1320–1326.
KNOX, K. W. (1966). *Biochem. J.*, **100**, 73–78.
KNOX, K. W., CULLEN, J., and WORK, E. (1967). *Biochem. J.*, **103**, 192–201.
KNOX, K. W., VESK, M., and WORK, E. (1966). *J. Bact.*, **92**, 1206–1217.
LANDY, M., and SHEAR, M. J. (1957). *J. exp. Med.*, **106**, 77–97.
LEIVE, L. (1965). *Biochem. biophys. Res. Commun.*, **21**, 290–296.
LEIVE, L., SHOVLIN, V. K., and MERGENHAGEN, S. E. (1968). *J. biol. Chem.*, **243**, 6384–6391.
LÜDERITZ, O. (1970). *Agnew. Chem.*, *int. edn*, **9**, 649–663.
LÜDERITZ, O., GALANOS, C., RISSE, H. J., RUSCHMANN, E., SCHLECHT, S., SCHMIDT, G., SCHULTE-HOLTHAUSEN, H., WHEAT, R., and WESTPHAL, O. (1966). *Ann. N.Y. Acad. Sci.*, **133**, 349–374.
LÜDERITZ, O., HAMMER, D., GOEBEL, F., SIEVERS, K., and WESTPHAL, O. (1958). *Z. Naturf.*, *B*, **13**, 566–571.

LÜDERITZ, O., JANN, K., and WHEAT, R. (1968). *Comprehensive Biochemistry*, **26A**, 105–228.
LÜDERITZ, O., STAUB, A. M., and WESTPHAL, O. (1966). *Bact. Rev.*, **30**, 192–255.
MCINTIRE, F. C., BARLOW, G. H., SIEVERT, H. W., FINLEY, R. A., and YOO, A. L. (1969). *Biochemistry, Wash.*, **8**, 4063–4067.
MCINTIRE, F. C., SIEVERT, H. W., BARLOW, G. H., FINLEY, R. A., and LEE, A. Y. (1967). *Biochemistry, Wash.*, **6**, 2363–2372.
MALCHOW, D., LÜDERITZ, O., KICKHÖFEN, B., and WESTPHAL, O. (1969). *Eur. J. Biochem.*, **7**, 239–246.
MARSH, D. G., and CRUTCHLEY, M. J. (1967). *J. gen. Microbiol.*, **47**, 405–420.
MARTIN, W. J., and MARCUS, S. (1966). *J. Bact.*, **91**, 1453–1459.
MARX, A., MUSETESCU, M., SENDREA, M., and MIHALCA, M. (1968). *Zentbl. Bakt. ParasitKde*, **207**, 313–316.
MILNER, K. C., and FINKELSTEIN, R. A. (1966). *J. infect. Dis.*, **116**, 529–536.
MORGAN, W. T. J. (1937). *Biochem. J.*, **31**, 2003–2021.
MORGAN, W. T. J., and PARTRIDGE, S. M. (1940). *Biochem. J.*, **34**, 169–191.
NETER, E., WESTPHAL, O., LÜDERITZ, O., GORZYNSKI, E. A., and EICHENBERGER, E. (1956). *J. Immun.*, **76**, 377–385.
NOLL, H., and BRAUDE, A. I. (1961). *J. clin. Invest.*, **40**, 1935–1951.
NOWOTNY, A. (1963*a*). *J. Bact.*, **85**, 427–435.
NOWOTNY, A. (1963*b*). *Nature, Lond.*, **197**, 721–722.
NOWOTNY, A. (1966). *Nature, Lond.*, **210**, 278–280.
NOWOTNY, A. (1969). *Bact. Rev.*, **33**, 72–98.
NOWOTNY, A., CUNDY, K. R., NEALE, N. L., NOWOTNY, A. M., RADVANY, R., THOMAS, S. P., and TRIPODI, D. J. (1966). *Ann. N.Y. Acad. Sci.*, **133**, 586–603.
NOWOTNY, A., KASAI, N., and TRIPODI, D. (1969). *Colloques int. Cent. natn. Rech. Scient.*, **174**, 79–94.
OSBORN, M. J., ROSEN, S. M., ROTHFIELD, L. and HORECKER, B. L. (1962). *Proc. natn. Acad. Sci. U.S.A.*, **48**, 1831–1838.
RAYNAUD, M., CHERMANN, J. C., and DIGEON, M. (1969). *Colloques int. Cent. natn. Rech. Scient.*, **174**, 47–67.
RAYNAUD, M., DIGEON, M. C., CHERMANN, J. C., and GIUNTINI, J. (1966). *C.r. hebd. Séanc. Acad. Sci., Paris*, **262**, ser. D, 722–724.
RAYNAUD, M., DIGEON, M., and NAUCIEL, C. (1964). In *Bacterial Endotoxins*, pp. 326–344, ed. Landy, M., and Braun, W. New Brunswick: Rutgers University Press.
RIBI, E., ANACKER, R. L., BROWN, R., HASKINS, W. T., MALMGREN, B., MILNER, K. C., and RUDBACH, J. A. (1966). *J. Bact.*, **92**, 1493–1509.
RIBI, E., ANACKER, R. L., FUKUSHI, K., HASKINS, W. T., LANDY, M., and MILNER, K. C. (1964). In *Bacterial Endotoxins*, pp. 16–28, ed. Landy, M., and Braun, W. New Brunswick: Rutgers University Press.
RIBI, E., HASKINS, W. T., LANDY, M., and MILNER, K. C. (1961). *J. exp. Med.*, **114**, 647–663.
RIBI, E., MILNER, K. C., and PERRINE, T. D. (1959). *J. Immun.*, **82**, 75–84.
RISSE, H. J., DROGE, W., RUSCHMANN, E., LÜDERITZ, O., and WESTPHAL, O. (1967). *Eur. J. Biochem.*, **1**, 216–232.
ROGERS, S. W., GILLELAND, H. E., and EAGON, R. G. (1969). *Can. J. Microbiol.*, **15**, 743–748.
ROTHFIELD, L., and PEARLMAN-KOTHENCZ, M. (1969). *J. molec. Biol.*, **44**, 472–492.
ROTHFIELD, L., and TAKESHITA, M. (1966). *Ann. N.Y. Acad. Sci.*, **133**, 384–390.
RUDBACH, R. A., and JOHNSON, A. G. (1966). *J. Bact.*, **92**, 892–898.
SHANDS, J. W. (1965). *J. Bact.*, **90**, 266–270.
SIEBERT, F. B. (1923). *Am. J. Physiol.*, **67**, 90–104.
SIEBERT, F. B. (1925). *Am. J. Physiol.*, **71**, 621–651.
SKARNES, R. C. (1966). *Ann. N.Y. Acad. Sci.*, **133**, 644–662.
SKARNES, R. C. (1968). *J. Bact.*, **95**, 2031–2034.
STETSON, C. A. (1956). *J. exp. Med.*, **104**, 921–933.

Taylor, A., Knox, K. W., and Work, E. (1966). *Biochem. J.*, **99**, 53–61.
Tripodi, D., and Nowotny, A. (1966). *Ann. N.Y. Acad. Sci.*, **133**, 604–621.
Watson, D. (1960). *J. exp. Med.*, **111**, 255–285.
Watson, D. W., and Kim, Y. B. (1963). *J. exp. Med.*, **118**, 425–446.
Weinstein, M. J., Waitz, J. A., and Came, P. E. (1970). *Nature, Lond.*, **226**, 170.
Westphal, O., Gmeiner, J., Lüderitz, O., Tanaka, A., and Eichenberger, E. (1969). *Colloques int. Cent. natn. Rech. Scient.*, **174**, 69–78.
Westphal, O., and Lüderitz, O. (1954). *Angew. Chem.*, **66**, 407–417.
Westphal, O., Lüderitz, O., and Bister, F. (1952). *Z. Naturf.*, *B*, **7**, 148–152.
Westphal, O., Lüderitz, O., Eichenberger, E., and Neter, E. (1955). *Dt. Z. Verdau.-u. StoffwechsKrankh.*, **15**, 170–180.
Westphal. O., Nowotny, A., Lüderitz, O., Hurni, H., Eichenberger, E., and Schönholzer, G. (1958). *Pharm. Acta Helv.*, **33**, 401–411.
Work, E. (1967). *Folia microbiol. Praha.*, **12**, 220–226.
Work, E. (1969). *Colloques int. Cent. natn. Rech. Scient.*, **174**, 35–45.
Work, E., Knox, K. W., and Vesk, M. (1966). *Ann. N.Y. Acad. Sci.*, **133**, 438–449.

DISCUSSION

Cranston: Are you certain that the organisms you are growing are all genetically identical?

Work: Yes, because it is a mutant which requires lysine.

Cranston: But couldn't other mutations arise within that group which also need lysine?

Work: Yes, they might.

Cranston: Do you start off with a single organism?

Work: Not every time. We go back to the single organism at intervals and we still get the same effects. We have been working with this organism now since 1959.

Palmer: You showed an increase in pyrogenic activity on treatment with alkali. How does that link up with Neter's work of exposing certain groups (Neter *et al.*, 1956)?

Work: This treatment (0·1N-NaOH, 18°C) was milder than Neter's (0·25N-NaOH, 56°C). It may tie up because Neter treated lipopolysaccharides with alkali to make them capable of coating erythrocytes to sensitize erythrocytes for antigenic tests. It is possible that a group may be exposed that will enable the molecule to do one thing (e.g. coat erythrocytes or promote endogenous pyrogen production) but after a little more treatment that group may be removed. There is, of course, a great difference between rate of loss of antigenic and pyrogenic activities in degraded lipopolysaccharides.

Palmer: Have you any idea which groupings are involved?

Work: I guess it is 2-keto-3-deoxyoctonic acid (KDO) or something joined to KDO, or it may be ethanolamine phosphate. KDO plays a great part in the aggregation and calcium binding of lipopolysaccharides.

Cooper: Have degraded portions of the lipopolysaccharide complex been tagged to see whether they will stick to white blood cells or other cells?

Work: No.

Whittet: The degradation by autoclaving is very interesting because the classical theory of pyrogens was that they were extremely thermostable. Dr Palmer and I have found that pyrogens range from being very thermostable to being very thermolabile, and we have even found some which have been destroyed by boiling for a short time. So the old theory that pyrogens are extremely thermostable is quite wrong (Whittet, 1958; Palmer, 1967; Palmer and Whittet, 1961).

Work: Our effect is very slow; you have to wait several weeks before you can spot it.

Whittet: There is a strong pyrogen in London tap-water which can be destroyed with the ordinary pharmacopoeial autoclaving process, and we have also found this with quite a number of purified pyrogens (Whittet, 1961; Palmer and Whittet, 1971).

REFERENCES

NETER, E., WESTPHAL, O., LÜDERITZ, O., GORZYRISKI, E. A., and EICHENBERGER, E. (1956). *J. Immun.*, **76**, 377–385.

PALMER, C. H. R. (1967). Ph.D. Thesis, University of London.

PALMER, C. H. R., and WHITTET, T. D. (1961). *J. Pharm. Pharmac.*, **13**, 62–66T.

PALMER, C. H. R., and WHITTET, T. D. (1971). *J. ind. Chem.*, in press.

WHITTET, T. D. (1958). Ph.D. Thesis, University of London.

WHITTET, T. D. (1961). *Publ. Pharm.*, **18**, 18–23.

THE SIGNIFICANT IMMUNOLOGICAL FEATURES OF BACTERIAL ENDOTOXINS

MAURICE LANDY

National Institute of Allergy and Infectious Diseases, Bethesda, Maryland

THE biology of the endotoxins of gram-negative bacteria (also referred to as somatic antigen and lipopolysaccharide) has been the subject of extensive study for decades. Most effort has been directed towards the extraordinary physiological and pharmacological derangements that these agents evoke in mammals. Their immunological attributes, while receiving later and less attention, are nonetheless distinctive in many ways. The basis for their serological specificity has been solidly related to the sequential and spatial conformation of their constituent sugars, but this facet of their reactivity is outside the scope of this report. For the purposes of this paper it is particularly useful to consider their extraordinary and diverse immunological characteristics in two separate categories: (1) the properties expressed when they are introduced parenterally (this is by far the most commonly used route); and (2) those properties and interactions that seem especially relevant to understanding the effects of endogenous gram-negative bacteria and endotoxin in the host; these properties are not necessarily related to their immunogenicity.

In considering the immunological performance of these antigens it is essential first to take into account the immunological status of the host as the test object. Quite unlike the spectrum of other antigens studied by the immunologist, these gram-negative bacterial components have exceedingly complex interrelationships with the mammalian species in which they are tested. To reduce the problem to manageable proportions I shall sub-divide host animals into three main categories: (*a*) virgin, (*b*) normal and (*c*) specifically sensitized.

(*a*) Even the so-called virgin host should really be considered in two separate classes: the germ-free colostrum-deprived ungulates, believed to be the only mammals free of demonstrable immunoglobulin (Sterzl *et al.*, 1965), and germ-free rodents—mice, rats, and guinea pigs—which, despite having no contact with bacteria, nonetheless have appreciable levels of natural antibodies reactive with endotoxins (Landy and Weidanz, 1964) and significant levels of immunoglobulins. Although the former lack immunoglobulins they still have responsive immunocompetent cells, whereas the latter have both immunoglobulins and immunocompetent cells. Both groups of animals are fully capable of responding immunologically to gram-negative bacteria, to their somatic antigens and to other antigenic stimuli. Indeed, colostrum-deprived piglets, which totally lack natural antibodies, show an enhanced immunological response by comparison with conventional (i.e. not "germ-free") animals. The germ-free rodent develops a natural background of immunity to bacterial antigens but with a somewhat lower level of antibody titres than conventional animals; however, this natural immunity can be stimulated in germ-free animals quite readily by administering bacterial antigens early in life and these animals thereby attain antibody levels comparable to those in conventional animals.

(*b*) The natural state of the gut in germ-free animals is quite unlike that of the normal host, which displays what some interpret as a chronic inflammatory state by comparison with the pale, comparatively non-lymphoid gut in the germ-free host. The normal host displays in his serum not only an extensive range of natural antibodies against all species of gram-negative bacteria but also antibodies directed against the erythrocytes and nucleated cells of virtually all other heterologous animal species (Terasaki *et al.*, 1961). Because of this all mammals should properly be regarded as having been sensitized to endotoxin by their environments, foodstuffs and bacteria. The gross and histological picture presented by the "normal" gut would seem to be consonant with such a view. In the so-called normal host the array of natural antibodies present are the predominantly high molecular weight (19S) pentavalent IgM immunoglobulins that are exceedingly efficient in complement-mediated bactericidal, viricidal and cytotoxic activities. When "normal" animals are first exposed parenterally to these somatic antigens, their response again consists of IgM antibodies and is very rapid. The immune response to these antigens, aside from its very rapid evolution and the predominance of the IgM type of antibody produced, is not strikingly different from the familiar pattern of response to protein antigens (Weidanz, Jackson and Landy, 1964), the rate and magnitude of the response being largely dependent on such factors as the form (particulate or soluble, high or low molecular weight) and nature of the antigen (much or little associated protein, high or low lipid content), its route of administration, the dose (nanogrammes or milligrammes) and the age of the animal (neonate or adult) (Landy, Sanderson and Jackson, 1965).

(*c*) A third category of host is the specifically sensitized animal in which, depending on the condition of immunization, the response is more likely to consist of low molecular weight (7S), divalent IgG immunoglobulin, as well as IgM and other immunoglobulin classes such as secretory IgA.

Unlike most antigens whose immunogenicity is powerfully enhanced by water-in-oil and mycobacterial adjuvants, the response to these bacterial components is quite unaffected by adjuvants. Indeed it was shown some 15 years ago (Johnson, Gaines and Landy, 1956) that endotoxin itself has powerful adjuvant activity when given together with any of a great variety of antigens.

Immunogenicity of bacterial endotoxins

While it is generally appreciated that these bacterial components are potent immunogens, two separate lines of work provide evidence that this immunogenicity is extraordinary, in that molecular amounts of these antigens, appropriately applied, produce a specific immune response. We (Landy and Baker, 1966) showed that injecting salmonella lipopolysaccharide into the foot pad of the rabbit evokes a rapid appearance of antibody-producing cells in the draining popliteal lymph node. The number of plaque-forming cells was an exquisitely sensitive indicator of the host's response to a single injection of salmonella lipopolysaccharide. At one extreme, these plaque-forming cells were elicited by as little as 1×10^{-9} μg, while the response was dose-related in the dose range 10^{-3} to 10^{-6} μg.

The long-held impression that these bacterial antigens do not evoke a secondary response has been refuted by Rudbach (1971), who demonstrated that, provided dosage and timing are appropriate, a marked secondary immune response can be obtained and that the dose of antigen which suffices to prime the animal for this effect is exceedingly small. Thus, purified *E. coli* somatic antigen in amounts ranging from 10^{-11} to 10^{-12} μg has been found adequate to prepare mice to give an accelerated, marked secondary response to a dose of 1 μg of the same antigen given two to three weeks later. These minute amounts of endotoxin correspond to about one to ten molecules of a 600 000 particle-weight element of endotoxin; this would represent 30–300 subunits. It is therefore clear that only a few molecules of this antigen are required to prime mice immunologically.

Of central importance in this issue of hyper-reponsiveness is the ubiquity of the natural antibodies directed against gram-negative bacteria. They appear so early in development as to preclude the possibility that they result from reactions to specific bacteria as such (Michael, Whitby and Landy, 1962). It is more likely that they reflect complex and subtle responses to environmental stimuli widely distributed in foodstuffs and therefore in continuous and intimate contact with the host. There is now good evidence that these are specific antibodies, as judged by quantitative absorption experiments (Michael, Whitby and Landy, 1962). They are present in all animal species studied, in all members of a species and their invariable presence has important consequences. This suggests that all "normal" animals are naturally sensitized to react to contact with bacterial antigens, whatever the source of the original stimulus, by continued production of these "natural" antibodies.

The somatic antigens or endotoxins extracted from Enterobacteriaceae have generally similar chemical and physical properties and can therefore be considered collectively. The refined materials are largely polysaccharide with variable amounts of associated bound lipid, and are contaminated with very small amounts of nucleic acids and proteins. They are usually extracted from bacterial cells by the phenol-water procedure and these extracts are sparingly soluble, polydispersed mixtures of endotoxin-polymers of molecular weights generally greater than one million. Ribi and co-workers (1966) have demonstrated that surfactants such as sodium deoxycholate lead to the dissociation of these polymers into subunits with molecular weights of only 10 000–20 000. These subunits are elongated molecules with dimensions approaching those of linear chains of hexoses; they retain the serological specificity of the original extract but are not immunogenic in experimental animals, nor do they evoke the characteristic physiological effects. If the surfactant is removed the subunits re-associate into the toxic, immunogenic polymers, with an average molecular weight ranging from 500 000 to 1 000 000.

The host possesses tissue components, notably in spleen, liver and the plasma, which on reaction with endotoxin *in vitro* also lead to its dissociation into subunits of 10 000–20 000 molecular weight (Chedid *et al.*, 1970). However, no covalent bonds are broken during this dissociation since, by appropriate procedures, the altered endotoxin can be restored to the original colloidal toxic components (Rudbach *et al.*, 1966). It has been demonstrated that in mice endotoxin is degraded *in vivo* and excreted in a form similar to this subunit. So far, no purified enzyme from any source, or for that matter any complex system from mammalian tissues, has been shown to *digest* and *destroy* endotoxin. Thus this relatively low molecular weight subunit appears to be the ultimate biological degradation product of endotoxin. Consequently antigen-processing cells or antibody-producing cells, whichever are the ultimate repository for these antigens, would be expected to contain a fragment of lipopolysaccharide with a molecular weight of at least 10 000–20 000. This is in contrast to protein antigens, which are made up of much smaller peptide fragments.

As information has accumulated on the immunological characteristics of these antigens it has become evident that they provide distinctive and valuable models for the basic study of the immune response itself. As with other major categories of antigens, marked differences in the immune response of different animal species have been seen; a considerable range of responses is also encountered in inbred lines within a species. A major genetic component must therefore determine the magnitude of the immune response to these lipopolysaccharides.

The persisting IgM antibody response to stimulation by endotoxin has been noted in all animal species examined. Moreover, unlike most other categories of antigenic responses, this one does not seem to involve thymus-dependent lymphocytes, as neonatal thymectomy has little if any effect on the production of this specific antibody. These antigens are thus capable of functioning without the cooperation of lymphocytes derived from the thymus, in contrast to protein antigens. The difference may relate to the predominance of the IgM, pentavalent type of antibody which endotoxins elicit. This type of antibody would be expected to be more effective in tightly binding, capturing and focusing this antigen for initiating the immune response than the smaller IgG (7S) antibody mainly elicited by protein antigens. This form of antibody is also very efficient and effective in bacterial lysis and bactericidal activities where it operates to the advantage of the host in resisting challenge by gram-negative bacterial pathogens.

A state of immunological unresponsiveness to bacterial somatic antigens, long considered to be unattainable because these components are such strong immunogens has been achieved by Friedman (1968). Single or multiple injections into neonatal mice render them specifically unresponsive to these antigens, as judged by markedly reduced amounts of circulating antibody and numbers of plaque-forming cells upon subsequent challenge. This immunosuppression is immunologically specific and has previously been obscured in some experiments when the challenge was in the form of intact bacteria, since they contain additional antigens which themselves induce an immune response when first introduced into the host. Endotoxin tolerance, long considered to be a non-specific response and not an immunologically based one, is now seen to be an example of immunological unresponsiveness. It can be passively transferred to other host animals by lymphoid cells, or by sera containing significant amounts of specific IgM antibody. The evidence as a whole points increasingly to the specific antibody as the essential requirement for this 'antipyrogen' immunity (Greisman, Young and Carozza, 1969).

INTERACTIONS BETWEEN ENDOTOXIN AND HOST

So far I have been principally concerned with the distinctive immunogenic attributes of endotoxin rather than with the *in vivo* consequences of these properties. I shall now try to identify some of the interactions and phenomena occurring in animal hosts that may be explicable in terms of the immunological properties of endotoxins so far identified.

Among the more important physiological changes that follow the administration of bacterial endotoxin are alterations of smooth muscle tone and vascular permeability. At their extreme, these events almost certainly contribute to the syndrome of endotoxin shock. The biochemical basis for these changes has remained obscure. The small amounts of endotoxin capable of bringing them about has, however, led to the surmise that endotoxin does not act directly upon tissue receptors, but rather acts indirectly by activating serum components such as those of the complement system. Recent investigations by Lichtenstein and co-workers (1969) and by Mergenhagen and co-workers (1970) on interactions between endotoxin and complement *in vitro* have suggested a possible mechanism by which this might happen. Endotoxin and complement exert dramatic effects on each other when they interact in fresh mammalian serum. Consumption of each of the terminal components of complement (C′3 to C′9) occurs, with minimal consumption of C′1, C′4 or C′2. This may be attributed to very efficient utilization of these early-acting components or possibly to an alternative pathway that does not involve these components. It is now established that activation of these terminal complement components is associated with the generation of several biologically active factors, one of which is anaphylatoxin, that could explain certain of the biological effects of endotoxin. Of paramount significance among these effects are changes in the clotting system, vascular permeability, smooth muscle reactivity and the chemotaxis of neutrophil leucocytes.

In the course of exploring the cellular aspects of the immune response of rabbits to salmonella lipopolysaccharide, we (Landy *et al.*, 1965) observed that within a few days after a single intravenous injection of 5 μg of this antigen, virtually all lymphocytes left the thymus and this organ, normally densely cellular, then became an empty bag. There was no histological evidence of cytotoxicity and further investigations of this phenomenon indicated that within two to three weeks the thymus once again became fully cellular, with normal histology. Later, Keast (1968) found that in neonatal mice endotoxin induced runting and pathological alterations that largely mimic the lesions of graft-versus-host responses occurring after the injection of foreign tissues, and also seen in mice that have been neonatally thymectomized. He found that a few injections of 10–15 μg of endotoxin during the first ten days of life assured that typical graft-versus-host reactions persisted, including thymic atrophy, which he attributed to a continuing leakage through the gut of endotoxin from the host's own bacterial flora. This type of damage could be prevented either by maintaining the mice in a germ-free environment or, in normal animals, by clearing the gut of gram-negative organisms by antibiotics, administered by gavage (Keast and Walters, 1968). Clearly, then, endotoxins exert a unique and striking effect on the thymus in normal animals and the thymic consequences are, if anything, markedly increased in situations that interfere with the host's normal antibacterial immunity.

Suppression of the immune response by either experimental or natural means generally renders the host markedly susceptible to the physiological effects of endotoxin, reflecting an incapacity to cope with the endogenous gram-negative flora. The outward expression is in runting or wasting disease, strikingly similar to that seen in thymectomized neonates and in animals undergoing graft-versus-host reactions to injected foreign cells, either as neonates or as X-irradiated adults. Other long-term consequences include the emergence of neoplasms of the lymphoreticular system and supervening gram-negative sepsis as a frequent cause of death.

CONCLUSIONS AND SUMMARY

The unique character of the interactions of the mammalian host with bacterial endotoxin is largely a reflection of the fact that all mammals have been naturally sensitized as a consequence of the ubiquity of gram-negative bacteria and their products. Moreover, the mammalian host carries in his gut, throughout his lifespan, the seed of his own destruction in the form of an enormous potential increase in the gram-negative flora of the gut. Normally this dangerous reservoir of endotoxin is held perfectly in check, presumably by the secretion by the gut of natural antibodies (IgA) that facilitate the sequestration and intracellular destruction of endotoxin. Any substantial emergence of endotoxin into the periphery is presumably dealt with by the plasma or tissue components that complex with and inactivate endotoxin *in vitro*. Alternatively, small amounts of endotoxin emerging into the circulation interact with natural antibody and complement components and thereby generate anaphylatoxin and other substances which in turn effect changes in the clotting system, vascular permeability, smooth muscle reactivity and in neutrophil chemotaxis. Immunosuppression of the host by antimetabolites or antilymphocyte serum, or by allogeneic cells results in graft-versus-host disease and a massive assault on the host by gram-negative bacteria and their products. Thus an intricate balance is normally maintained between the physiological threat represented by the host's endogenous supply of endotoxin and an array of immunological systems that constitute a dynamic defence.

REFERENCES

CHEDID, L., PARANT, M., PARANT, F., and PEREZ, J. J. (1970). *Infect. Immunity*, **1**, 15–20.
FRIEDMAN, H. (1968). *J. Bact.*, **96**, 1124–1132.
GREISMAN, S. E., YOUNG, E. J., and CAROZZA, F. A. (1969). *J. Immun.*, **103**, 1223–1236.
JOHNSON, A. G., GAINES, S., and LANDY, M. (1956). *J. exp. Med.*, **103**, 225–246.
KEAST, D. (1968). *Immunology*, **15**, 237–245.
KEAST, D., and WALTERS, M. N. (1968). *Immunology*, **15**, 247–262.
LANDY, M., and BAKER, P. J. (1966). *J. Immun.*, **97**, 670–679.
LANDY, M., SANDERSON, R. P., BERNSTEIN, M. T., and LERNER, E. M. (1965). *Science*, **147**, 1591–1592.
LANDY, M., SANDERSON, R., and JACKSON, A. L. (1965). *J. exp. Med.*, **122**, 483–504.
LANDY, M., and WEIDANZ, W. P. (1964). In *Bacterial Endotoxins*, pp. 275–290, ed. Landy, M., and Braun, W. New Brunswick: Rutgers University Press.
LICHTENSTEIN, L. M., GEWURZ, H., ADKINSON, N. F., SHIN, H. S., and MERGENHAGEN, S. E. (1969). *Immunology*, **16**, 327–336.
MERGENHAGEN, S. E., SNYDERMAN, R., GEWURZ, H., and SHIN, H. S. (1970). *Curr. Top. Microbiol.*, **49**, 1–35.
MICHAEL, J. G., WHITBY, J. L., and LANDY, M. (1962). *J. exp. Med.*, **115**, 131–146.
RIBI, E., ANACKER, R. L., BROWN, R., HASKINS, W. T., MALMGREN, B., MILNER, K. C., and RUDBACH, J. A. (1966). *J. Bact.*, **92**, 1493–1507.
RUDBACH, J. A. (1971). *J. Immun.*, in press.
RUDBACH, J. A., ANACKER, R. L., HASKINS, W. T., JOHNSON, A. G., MILNER, K. C., and RIBI, E. (1966). *Ann. N.Y. Acad. Sci.*, **133**, 629–643.
STERZL, J., MANDEL, L., MILER, J., and RIHA, I. (1965). In *Molecular and Cellular Basis of Antibody Formation*, pp. 351–370, ed. STERZL, J. Prague: Czechoslovak Academy of Sciences.
TERASAKI, P. I., ESAIL, M. L., CANNON, J. A., and LONGMIRE, W. P. (1961). *J. Immun.*, **86**, 383–395.
WEIDANZ, W. P., JACKSON, A. L., and LANDY, M. (1964). *Proc. Soc. exp. Biol. Med.*, **116**, 832–837.

DISCUSSION

Atkins: Rather than modifying the immunological defence mechanisms specifically, a large dose of endotoxin might produce shock, causing the affected animal to develop bacteraemia and die.

Landy: This is possible. However, although our animals reacted to the toxicity of the 5 μg dose, very few succumbed. At the time we were well aware that the lymphoid alterations in the thymus (elimination of thymocytes) (Landy *et al.*, 1965) might be induced via corticoids released from the adrenals, as it is known that corticoids can also cause some of these effects. However, reasons were given why this mechanism did not seem likely. Our conclusion was that the appearance of specific antibody-forming cells in this organ reflected a free and uninhibited traffic of such cells (Landy, Sanderson and Jackson, 1965), though antigen as such does not normally penetrate the thymus.

Bondy: Can an animal which has never been exposed to bacterial antigens release leucocyte pyrogen in response to an initial stimulus to which it has not previously been exposed?

Landy: I am not certain that such a precise, controlled experiment has ever been done. We should consider separating the several categories of test animals that are available even in the germ-free state; the rodent gnotobiotes, for example, have immunoglobulins, implying some prior contact with antigens, presumably via the diet which is germ-free but not antigen-free. For definitive experiments it would be desirable to work with immunoglobulin-free animals, such as colostrum-deprived piglets, since it is a reasonable assumption that they are pristine and unsullied by prior sensitization.

Atkins: Germ-free animals get endotoxins in their food, so it is much harder to produce an endotoxin-free animal, in the sense implied here, than a germ-free animal.

Landy: It is, however, possible. The group at the Lobund Institute are able to maintain rats on a diet consisting exclusively of low molecular weight components, that is amino acids, minerals and vitamins.

Cooper: It is known that some bacterial pyrogens can produce fever if administered in an aerosol and that some infections in the lung can also give rise to fever. Could the surface-active material lining the lung alveoli be one means whereby this rather large molecule is assisted into the blood stream?

Landy: Some of my colleagues developed marked respiratory sensitivities after handling large quantities of dried gram-negative bacteria in the course of pulverizing these materials and in situations where aerosols might have been created. There is an entire literature of clinical syndromes that involve such airborne exposure to bacterial endotoxins. These reports indicate that there are, in nature, counterparts of this laboratory experience.

Bondy: We should not underestimate the importance of the lung macrophages which may be very important in picking up the bacterial pyrogens.

Work: Westphal and co-workers (1969) found that glycolipids (composed of lipid A + KDO), which are only slightly soluble in water, had their pyrogenicity increased tenfold when dispersed in triethylamine, showing that the degree of dispersion of pyrogens can be one factor determining biological activity.

REFERENCES

LANDY, M., SANDERSON, R. P., BERNSTEIN, M. T., and LERNER, E. M. (1965). *Science*, **147**, 1591–1592.

LANDY, M., SANDERSON, R. P., and JACKSON, A. L. (1965). *J. exp. Med.*, **122**, 483–504.

WESTPHAL, O., GMEINER, J., LÜDERITZ, O., TANAKA, A., and EICHENBERGER, E. (1969). *Colloques int. Cent. natn. Rech. Scient.*, **174**, 69–78.

PURIFICATION OF AN ENDOGENOUS PYROGEN, WITH AN APPENDIX ON ASSAY METHODS*

P. A. MURPHY, P. J. CHESNEY and W. B. WOOD, JR

Department of Microbiology, The Johns Hopkins University School of Medicine, Baltimore, Maryland

IN the last century it was noticed that fever sometimes followed the intravenous injection of a large variety of apparently unrelated substances. Following the work of Hort and Penfold (1912) and Siebert (1923) it became accepted that all these fevers resulted from contamination of the materials by bacterial endotoxins. The first worker to establish the independent existence of endogenous pyrogens was Beeson (1948), who showed that saline extracts of rabbit granulocytes contained a substance which caused fever when injected intravenously into rabbits. Bennett and Beeson (1953) confirmed these findings and showed that extracts of various other tissues were not pyrogenic unless an infiltrate of polymorphonuclear leucocytes was present. "Leucocyte pyrogen" was a heat-labile protein whose biological properties differed from those of endotoxin. Subsequently, a series of papers from Wood's laboratory (summarized by Wood, 1958) established that leucocyte pyrogen was the common effector substance for a variety of experimental fevers, and Bennett (1956) showed that it seemed to be responsible for the fevers which develop during pneumococcal infections in rabbits.

Leucocyte pyrogen could conveniently be obtained from rabbit peritoneal exudate cells by incubating them in saline at 38°C (King and Wood, 1958). It could also be obtained from blood cells, but these had to be stimulated to produce it by bacterial endotoxin (Cranston *et al.*, 1956), phagocytosis (Berlin and Wood, 1964) or other techniques.

The evidence that leucocyte pyrogen was not a fragment derived from bacterial endotoxins was all circumstantial because the molecule was not chemically defined. Several attempts to purify it were made, the most successful being that of Kozak and co-workers (1968) who obtained a protein from acrylamide gels which caused fevers of 0·6–0·8°C when injected into rabbits in a dose of 1 μg. This protein appeared to be homogeneous when electrophoresed at pH 9 or at pH 3·5, and it was tentatively suggested that pure leucocyte pyrogen had been obtained.

* Dr Murphy was unable to attend and his paper was presented by Dr Snell.

However, experiments (Murphy, Neidengard and Wood, 1971) using acrylamide gels of smaller pore size than those used by Kozak and co-workers showed clearly that the band which had been thought to be homogeneous was composed of several proteins, and furthermore that leucocyte pyrogen could not be identified with any of the visible bands. This suggested that the pyrogen had a specific activity very much greater than had been supposed, and that it constituted only a minute fraction of the proteins present in the mixture. It was found that pyrogen isolated from acrylamide gels was too unstable to work with, so we developed new methods of purifying it. We are now able to obtain leucocyte pyrogen which appears to be homogeneous and which causes average fevers of 1°C when injected in doses of less than 0·1 μg protein.

Crude pyrogen from rabbit peritoneal exudate cells contained much fibrinous debris and DNA, and the first stage was to filter it through a membrane (XM-100 Membranes, Amicon Corporation, Scientific Systems Division, Lexington, Massachusetts, 02173, U.S.A.) which retained molecules of molecular weight greater than 100 000, but permitted smaller ones to pass. It was found that high yields of pyrogen could only be obtained by conducting the filtration at pH 3·5–3·7. At this pH the immediate yield of pyrogen was approximately 100 per cent, but it had a pronounced tendency to deteriorate on storage. Changing the pH did not improve matters and, since much purer specimens showed no tendency to deteriorate, we assumed that the pyrogen was being inactivated by one of the other proteins in the filtrate. The deterioration could be minimized by storing the concentrated filtered pyrogen in liquid nitrogen.

The filtrates from ten litres of crude pyrogen were pooled and concentrated to 100–120 ml. This concentrate was chromatographed on Sephadex G-50 (Pharmacia A.B., Uppsala, Sweden), using 0·2M-sodium acetate buffer, pH 4·2. The distributions of protein and pyrogen in the effluent are shown in Fig. 1. There was no correlation between the pyrogenic activity and any of the visible protein peaks, suggesting that pyrogen constituted only a minute proportion of the proteins in the active fractions. The pyrogen peak was closely followed by a large unrelated protein peak from which it could be resolved only when the sample size was 1·5 per cent of the column volume, or less.

All fractions which contained pyrogen of specific activity 8 μg protein/°C fever, or better, were accepted for further processing, The edges of the distribution were retained and re-run on a smaller G-50 column. The fractions of high specific activity were pooled with those of the initial run.

The overall yield of this process averaged 80 per cent and about 5 μg protein caused a mean fever of 1°C in rabbits. Material of higher specific activity, 1–2 μg/°C fever, could be obtained by re-running the concentrated fractions on a second G-50 column. This did not offer any obvious advantages in further processing, and the losses incurred were substantial so we accepted the less pure product obtained by one run. However, it was interesting to note that the pyrogen obtained by two gel filtrations had a specific activity at least equal to that of the protein isolated by Kozak and her co-workers (1968).

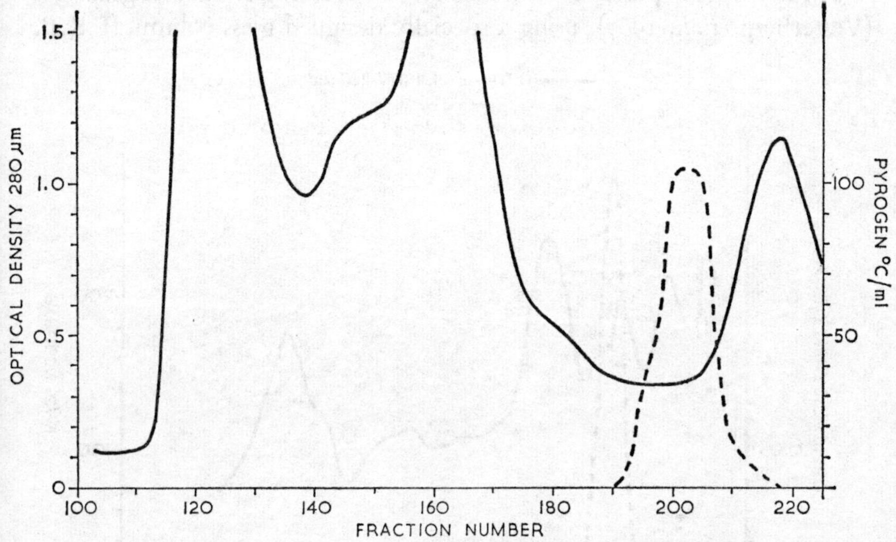

FIG. 1. Distribution of protein (optical density 280) (——) and pyrogen (---) in the effluent from a Sephadex G-50 column. Column size 10 × 100 cm, sample size 100 ml, sodium acetate buffer, 0·2 M, pH 4·2, flow rate 250 ml/hour.

The pyrogen obtained by gel filtration was subjected to ion-exchange chromatography on QAE Sephadex (Pharmacia A.B., Uppsala, Sweden), a substituted dextran carrying quaternary ammonium groups. Tris-hydrochloric acid buffer, 0·01M, pH 8·1, was used to apply the sample, and a gradient to 0·1M buffer was applied to elute the pyrogen. Again there was no good correlation between pyrogenic activity and an optical density peak (Fig. 2), suggesting that little of the measured optical density was due to pyrogen. There was clear evidence that pyrogen was heterogeneous; at least two long low peaks of pyrogenic activity followed the main pyrogen peak. When mercaptoethanol was added to all fractions a fourth peak appeared straddling the second and third. Leucocyte pyrogen was known to become oxidized and inactivated at alkaline pH (Kozak *et al.*, 1968), and we assumed that the fourth peak represented oxidized leucocyte pyrogen.

The specific activity of pyrogen eluted from ion-exchange columns varied between 0·1 and 0·3 μg protein/°C fever. Acrylamide gel electrophoresis still showed several protein bands, but it was possible to identify the pyrogenic activity with one of them. In our early work we could not identify pyrogen with a visible band, but we were now processing much larger quantities of material and, in addition, were staining the gels with Coomassie Blue instead of with the less sensitive Amido-Schwarz reagent.

The last stage in purification was isoelectric focusing on sucrose gradients (Vesterberg *et al.*, 1967), using a specially designed glass column (L.K.B.

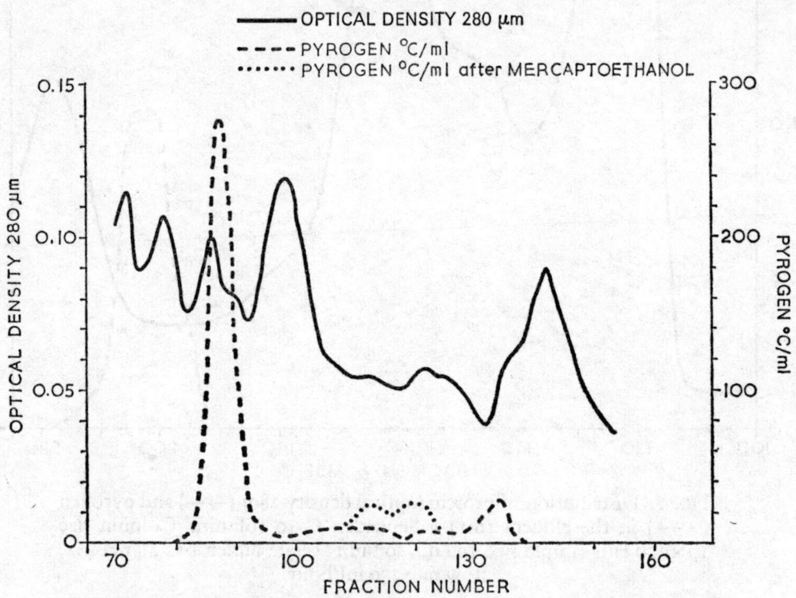

FIG. 2. Distribution of protein (optical density 280) and pyrogen in the effluent from a QAE Sephadex column. Column size 2·5 × 20 cm, sample size 30 ml, tris/HCl buffer, 0·01 M—0·1 M, pH 8·1, flow rate 10 ml/hour.

Produkter A.B., Stockholm-Bromma, Sweden). We had a good deal of trouble with this technique because of precipitation of pyrogen at its isoelectric point and because of its tendency to become oxidized and inactivated. To prevent these difficulties it was found necessary to conduct the focusing in the presence of 1M-urea, 15 per cent dimethylformamide and 2 per cent mercaptoethanol. In these conditions the isoelectric point of the main component shifted from 7·2 to 7·8, which presumably reflected partial unfolding of the protein chain. However, the biological activity was unaffected by this treatment.

We divided the pyrogen from the ion-exchange column into two parts; the main peak was focused alone in one experiment, and all the subsequent peaks were focused together in a second. The main peak focused as a single pyrogen peak with maximum activity at pH 7·78 (Fig. 3). The overall yield of this step was 80 per cent, and the two peak tubes contained three-quarters of that. Samples of the peak were dialysed to remove the low molecular weight solutes and were subjected to electrophoresis to determine the number of protein bands they contained. Electrophoresis at pH 9 was performed using the buffer system of Davis (1964) with the resolving gel modified to contain 15 per cent acrylamide. Electrophoresis

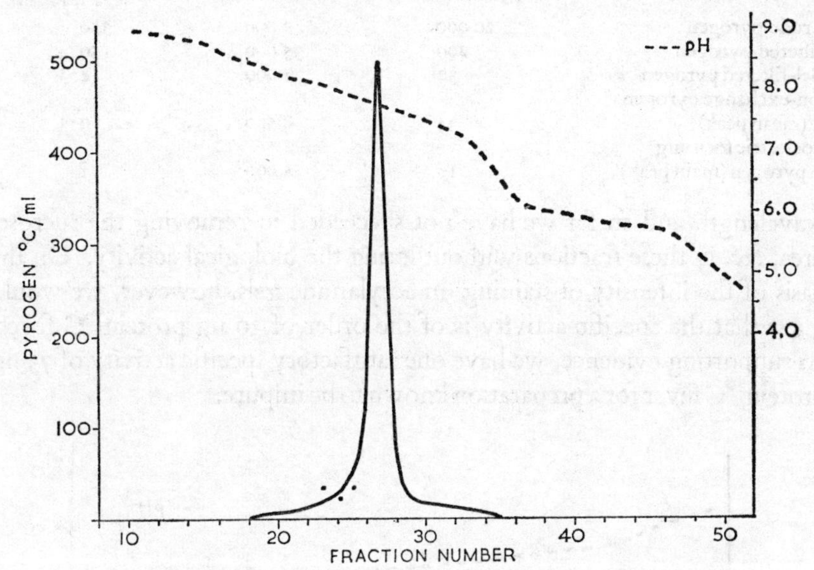

FIG. 3. Distribution of pyrogen in the effluent from an isoelectric focusing column. Sample: main pyrogen peak from QAE Sephadex chromatography. Column volume 440 ml, 1% ampholytes, pH range 7–9, final voltage 1200, current 3 mA, focusing time 4 days.

at pH 3·5 was conducted using a 15 per cent acrylamide gel and 0·025M-potassium acetate, pH 3·5. No stacking or buffer discontinuities were used. A third sample was subjected to isoelectric focusing in acrylamide gel, using a modified version of the technique of Catsimpoolas (1968).

In all three systems, only one protein band was visible on the stained gels. At pH 3·5 and at pH 9 we established that the pyrogenic activity coincided with the protein band seen on a stained section of the same gel. However, it should be noted that the homogeneity of the pyrogen was tested by methods which depend on the same molecular properties as the methods used to purify it. The evidence would be much stronger if an antibody against the pyrogen could be found, and if homogeneity were demonstrated by double diffusion in agar, or by immunoelectrophoresis.

We have not yet been able to obtain the specific activity of our final preparations of leucocyte pyrogen. The quantity of protein in our most active fractions is so small that the only technique which is sufficiently sensitive to measure it is the optical density at 191 nm (Mayer and Miller, 1970). Almost any solute other than sodium fluoride absorbs light at this

TABLE I

PURIFICATION PROCESS FOR THE MAIN COMPONENT OF LEUCOCYTE PYROGEN

Material	*Volume* (ml)	*Pyrogen content* (°C)	*Specific activity* (μg/°C)
Crude pyrogen	20 000	40 000	350
Filtered pyrogen	200	35 000	150
Gel-filtered pyrogen	32	28 000	4·5
Ion-exchange pyrogen (mean peak)	34	8 500	0·3
Isoelectric focusing pyrogen (main peak)	16	5 000	?

wavelength and so far we have not succeeded in removing the sucrose, urea, etc. in these fractions without losing the biological activity. On the basis of the intensity of staining on acrylamide gels, however, we would guess that the specific activity is of the order of 30 ng protein/°C fever. As supporting evidence, we have one satisfactory specific activity of 73 ng protein/°C fever for a preparation known to be impure.

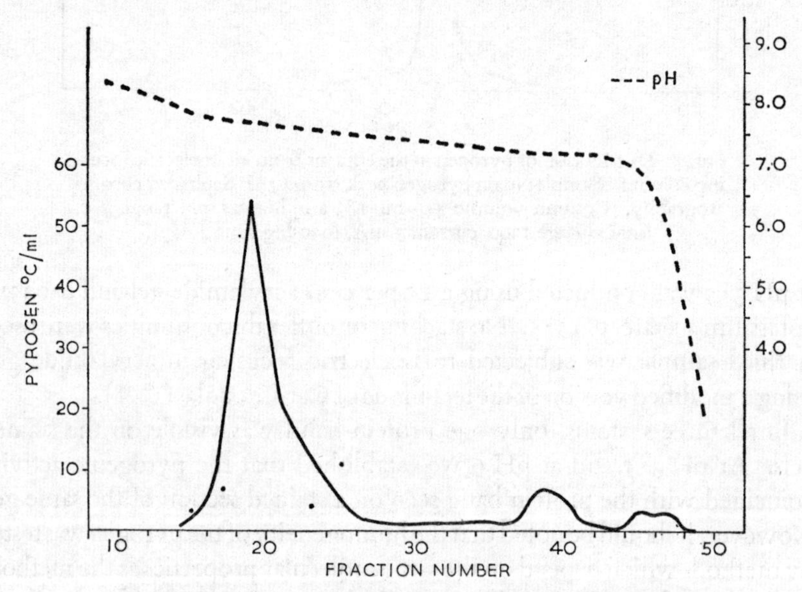

FIG. 4. Distribution of pyrogen in the effluent from an isoelectric focusing column. Sample: minor peaks from QAE Sephadex chromatography. Column volume 440 ml, 1% ampholytes, pH range 7–9, final voltage 1200, current 3 mA, focusing time 4 days.

The overall purification process is set out in Table I, together with yields and specific activities. It is possible that we may eventually be able to improve on the final yield, since most of the losses seem to occur during storage in the intermediate stages. Pyrogen in the effluent from the isoelectric focusing columns is quite stable.

The results of ion-exchange chromatography suggested that crude pyrogen was heterogeneous. This was confirmed by pooling the three later peaks and subjecting them to isoelectric focusing in sucrose. They segregated into three clear peaks (Fig. 4) of which the first was much the largest and was identical to the single pyrogen peak seen in the first experiment. This peak accounts for at least 90 per cent of the pyrogenicity of crude material. The minor components might be secreted by other cells, such as monocytes, or could conceivably represent degradation products of the molecule in the main peak. It would be interesting to look at the chemical properties of pyrogens obtained from pure populations of cells.

APPENDIX

STATISTICAL ASPECTS OF THE ASSAY OF PYROGENS

As far as we know, the pyrogen molecule has no special chemical or physical property which is unique to it and could be made the basis of an assay procedure. Thus we can only measure the pyrogen content of solutions by inference from the fevers which they cause when injected into rabbits. No biological assay has any absolute validity; always we must express the pyrogenicity of one solution as a percentage of that of some other solution. There is no international standard leucocyte pyrogen with which experimental materials can be compared. This means that it is not possible to publish figures for specific activity which could readily be checked in other laboratories by reference to the standard material. Because of the length of time (3–4 hours) which had to elapse between successive injections it was only possible to conduct assays in statistically correct fashion when we wished to compare the pyrogenicity of one solution with that of another. In this situation we performed parallel line assays as described by Finney(1950); an example is the determination of the yield of pyrogen during the filtration process which was the first step in purification.

The principle was to construct short log-dose response lines for both crude and filtered pyrogens using three doses of each in the ratio 1:2:4

TABLE II

SIX POINT COMPARISON BETWEEN CRUDE AND FILTERED PYROGEN

	Crude pyrogen				*Filtrate*		
Dose (ml)	*Fevers (°C)*	*Total*	*Mean (°C)*	*Dose (ml)*	*Fevers (°C)*	*Total*	*Mean (°C)*
3/16	0·35, 0·3, 0·25, 0·25, 1·1, 0·5, 0·5, 0·1	3·35 (8)	0·42	9/32	0·65, 0·35, 0·3, 0·15, 0·45, 0·45, 1·05, 0·75	4·15 (8)	0·52
3/8	1·05, 0·5, 0·7, 0·8, 0·55, 1·0, 1·05, 0·65	6·30 (8)	0·79	9/16	0·5, 0·65, 0·9, 0·5, 1·2, 1·15, 0·9, 0·4	6·20 (8)	0·78
3/4	1·1, 0·6, 0·45, 0·7, 0·8, 1·0, 2·1, 1·25	8·0 (8)	1·00	9/8	0·85, 1·0, 0·8, 1·05, 1·25, 1·35, 1·55, 0·9	8·75 (8)	1·09
Volume	2000 ml			Volume	2900 ml		

(Fig. 5). Parallel lines were fitted to the points as shown. Since the lines were parallel, the horizontal distance, M, between them was constant. The anti-logarithm of M was the ratio of doses of crude and filtered pyrogen which cause equal febrile responses. M could be obtained either from the graph, using lines fitted by eye, or by calculation from the data. The calculations involved were tedious and most of the information could be obtained directly from the data by simple inspection (Table II). The total volume of filtrate was about 1·5 times that of the crude pyrogen, and each

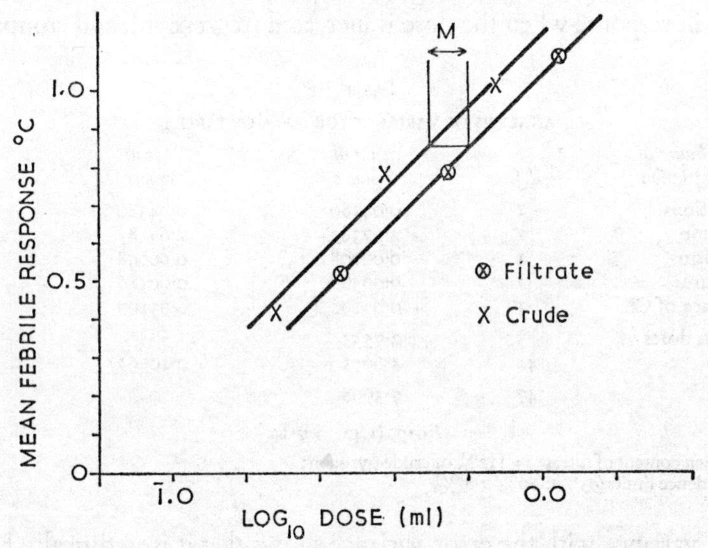

FIG. 5. Six-point parallel line comparison of crude pyrogen and filtered pyrogen.

dose of filtrate was 1·5 times that of the crude pyrogen. If the yield were 100 per cent, therefore, equal fevers should be produced by corresponding doses of the preparations, and it can be seen that this is in fact so. If the response to the lowest dose of crude pyrogen had been equal to that of the middle dose of filtrate, the yield would have been 50 per cent, and by interpolation intermediate values could be guessed with fair accuracy.

The importance of conducting assays in this way was that confidence limits could be assigned to the answer obtained. In this case the filtrate was calculated to contain 112 per cent of the pyrogenicity of the crude material. The 95 per cent confidence limits were 80 and 199 per cent, which shows how inaccurate these assays can be, despite the greatest care, and underlines how useless is quantitative information based on the responses of a few rabbits to one dose of a preparation. Multiple assays giving closely similar results are the only safeguard against error.

The second reason for doing parallel line assays was that the assumptions underlying the assay could be checked by analysis of variance, using the data of the experiment (Table III). Animals are not machines, and their responses do not always conform to the mathematical formulae we use to describe them. The total variance of all 48 responses in Table II could be split into two components. One of these was the error variance, the mean variance for the response of rabbits to *one* dose of pyrogen. The difference between the error variance and the total variance can be split into five separate components. The largest component is the variance due to increase in response when the dose is increased (regression), and comparison

TABLE III

ANALYSIS OF VARIANCE FOR DATA OF TABLE II

Nature of variation	*d.f.*	*Sum of squares*	*Mean square*	*F*
Preparations	1	0·04380	0·04380	<1
Regression	1	2·67383	2·67383	24·4
Parallelism	1	0·00008	0·00008	<1
Curvature	1	0·00586	0·00586	<1
Difference of C.	1	0·03190	0·03190	<1
Between doses	5	2·7555		
Error	42	4·6053	0·10965	
Total	47	7·3608		

$F_{0\cdot 99}(1,42) = 7\cdot 27$

Pyrogen content of filtrate = 112% of crude pyrogen;
Confidence limits (95%): 80%, 199%.

of this variance with the error variance shows that it is statistically highly significant. If it were not, the whole assay would be invalid, for it would not be possible to assert that the dose of pyrogen given had any effect on the response observed. The variance for difference between the preparations was not significant, but it had to be related to the doses given before any quantitative conclusion could be drawn. In this case, corresponding doses were equal fractions of the total pool of material, so that the non-significance of the variance of the preparations showed that the confidence limits of the pyrogen ratio would include 100 per cent. The other components measured deviations of the experimental points from the ideal parallel straight lines. Thus the lines might be straight, but not parallel; they might be curved, and if curved they might be curved to a different degree. The variances for these components were all much smaller than the error variance, so the data conformed well to the theoretical expectations, and the results of the assay could therefore be regarded as valid.

In routine practice, we used a four-point assay procedure. With this design curvature cannot be tested separately from parallelism, but this was

TABLE IV

FOUR-POINT COMPARISON BETWEEN FILTERED PYROGEN AND SEPHADEX PURIFIED PYROGEN

Dose (μl)	*Fevers* (°C)	*Total*	*Mean* (°C)
2·5	0·4, 0·65, 0·08, 0·25, 0·37, 0·8, 0·7, 0·46	3·71 (8)	0·46
5	0·7, 1·4, 0·63, 1·04, 0·97, 0·78, 0·98, 0·72	7·22 (8)	0·90
Volume	22 ml		

Dose (μl)	*Fevers* (°C)	*Total*	*Mean* (°C)
10	0·1, 0·52, 0·43, 0·3, 0·65, 0·58, 0·25, 0·05	2·88 (8)	0·36
20	0·6, 1·18, 0·4, 0·75, 0·72, 0·68, 0·92, 0·78	6·03 (8)	0·75
Volume	95 ml		

not thought to be important enough to justify the extra time and trouble involved in giving two extra doses. The graphical analysis was exactly as for the six-point assay (Fig. 6); the results could be read off approximately by inspection of the data (Table IV) and the validity of the assay could be checked by analysis of variance (Table V).

The specific activities which we have quoted were obtained by constructing a short log dose-response line for the preparation in question and calculating the point at which the line crossed 1°C. The protein content of the preparation was determined, and the specific activity defined as that amount of protein which produced a mean fever of 1°C in rabbits. It must be admitted that very serious errors were introduced by this

TABLE V

ANALYSIS OF VARIANCE FOR DATA OF TABLE IV

Nature of variation	*d.f.*	*Sum of squares*	*Mean square*	*F*
Preparations	1	0·1275	0·1275	2·29
Regression	1	1·3861	1·3861	24·9
Parallelism	1	0·00405	0·00405	<1
Between doses	3	1·5177		
Error	28	1·5577	0·05563	
Total	31	3·0754		

$F_{0.95}(1,28)=4.2$; $F_{0.99}(1,28)=7.64$

Pyrogen content of Sephadex-purified pyrogen=87·5% that of filtrate; 95% confidence limits: 60%, 117%.

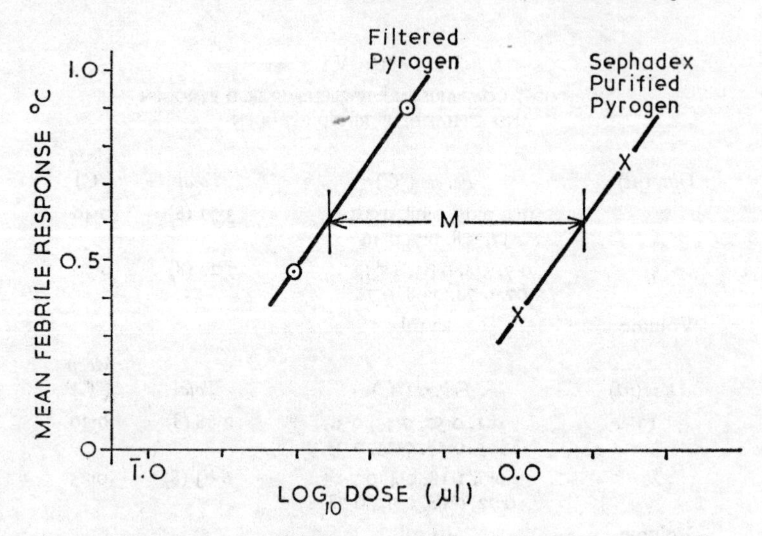

FIG. 6. Four-point parallel line comparison of filtered pyrogen and Sephadex purified pyrogen.

technique, since the response of rabbits to pyrogens varies from day to day. Nevertheless, when the same group of assay rabbits was used for long periods reasonably consistent values were observed. These values were only

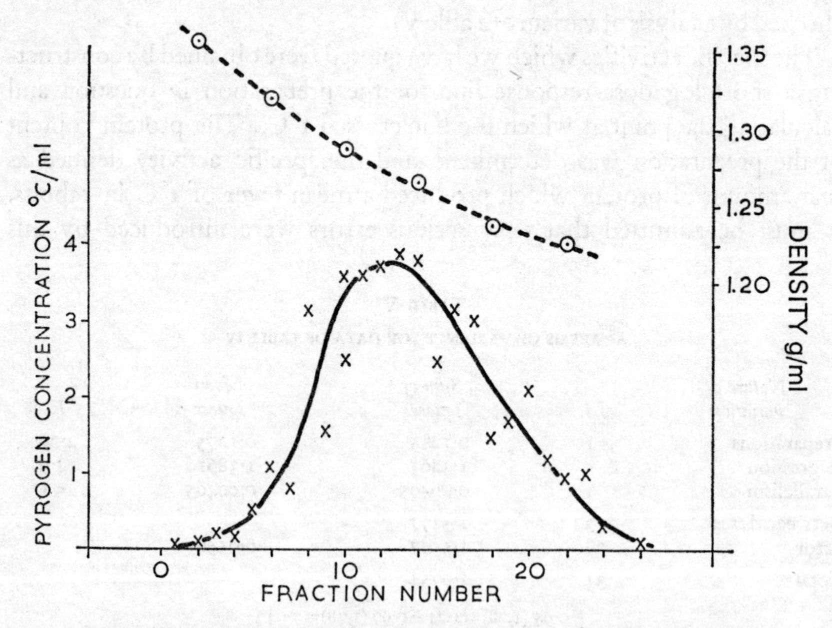

FIG. 7. Distribution of pyrogenic activity on a caesium chloride gradient.

applicable in our laboratory, for each pyrogen worker uses rabbits from different sources, and selects rabbits for assay work using his own arbitrary criteria.

The distribution of pyrogen in column effluents was determined by even less satisfactory methods. It was impractical to conduct parallel line assays on twenty or thirty fractions, so we used only one dose of each fraction tested. We worked in from the edges of the distribution, testing in a dose at least ten times that which would cause a substantial fever at the peak. When febrile responses started to occur we progressively reduced the dose given to keep the responses at or below 1 °C. Tabulation of the results gave a crude but serviceable definition of the peak.

It was convenient to have some idea of the specific activity of the fractions; this could only be done by guesswork, but with practice the guesses were approximately correct. From the measured response to a dose of the fraction (usually in only four rabbits) we inferred the dose of that fraction which would have caused a fever of 1°C. This was done graphically, reading off from a regression line with a slope of 0·3°C increase in fever for a doubling of dose. We could then calculate the concentration of pyrogen in °C/ml, and the specific activity followed from the measured protein concentration.

The errors in the values obtained for any one tube were, of course, astronomical. But the results from adjacent tubes were mutually supporting, and it was usually possible to draw a sensible curve through them. As evidence of this, we offer an experiment determining the buoyant density of pyrogen on a caesium chloride gradient (Fig. 7). Each fraction was tested in four rabbits, and the pyrogen concentration in °C/ml calculated as described. The points show considerable scatter, but it is clear that the buoyant density was 1·26 or 1·27 which was accurate enough for our purposes. A second experiment gave exactly the same result.

SUMMARY

Leucocyte pyrogen obtained from rabbit peritoneal exudate cells has been purified until it appears to be homogeneous. The specific activity of the crude material was such that 300–400 μg of protein caused a mean fever of 1°C in rabbits. Crude pyrogen was filtered through Amicon XM-100 membranes at acid pH, and the filtrate chromatographed on Sephadex G-50. The best material so obtained had a specific activity of 1–2 μg/°C fever. Chromatography on QAE Sephadex or CM Sephadex gave material with an activity of 0·1–0·2 μg/°C. It was still separable into several bands on acrylamide gel electrophoresis. Homogeneous pyrogen could be obtained by extracting the active band from such gels or by isoelectric focusing on sucrose gradients. Homogeneity was demonstrated by acrylamide gel electrophoresis at pH 9 and pH 3·5 and by isoelectric focusing in acrylamide gel. The purified pyrogen is highly unstable and so far its specific activity has not been measured satisfactorily. As judged by intensity of staining on acrylamide, its activity seems to be 20–30 ng/°C fever.

REFERENCES

BEESON, P. B. (1948). *J. clin. Invest.*, **27**, 524.

BENNETT, I. L. JR (1956). *Bull. Johns Hopkins Hosp.*, **98**, 216–235.

BENNETT, I. L. JR, and BEESON, P. B. (1953). *J. exp. Med.*, **98**, 477–491, 493–508.

BERLIN, R. D., and WOOD, W. B. JR (1964). *J. exp. Med.*, **119**, 715–726.

CATSIMPOOLAS, N. (1968). *Analyt. Biochem.*, **26**, 480–482.

CRANSTON, W. I., GOODALE, F. JR, SNELL, E. S., and WENDT, F. (1956). *Clin. Sci.*, **15**, 219–226.

DAVIS, B. J. (1964). *Ann. N.Y. Acad. Sci.*, **121**, 404–427.

FINNEY, D. J. (1950). In *Biological Standardisation*, 2nd edn, pp. 72–94, by Burn, J. H., Finney, D. J., and Goodwin, L. G. London: Oxford University Press.

HORT, E. C., and PENFOLD, W. J. (1912). *J. Hyg., Camb.*, **12**, 361–390.

KING, M. K., and WOOD, W. B. JR (1958). *J. exp. Med.*, **107**, 279–303.

KOZAK, M. S., HAHN, H. H., LENNARZ, W. J., and WOOD, W. B. JR (1968). *J. exp. Med.*, **127**, 341–357.

MAYER, M. M., and MILLER, J. (1970). *Analyt. Biochem.*, **36**, 91–100.

MURPHY, P. A., NEIDENGARD, L., and WOOD, W. B. JR (1971). In press.

SIEBERT, F. (1923). *Am. J. Physiol.*, **71**, 621–651.

VESTERBERG, O., WADSTROM, T., VESTERBERG, K., and SVENSSON, H. (1967). *Biochem. biophys. Acta*, **133**, 435–445.

WOOD, W. B. JR (1958). *Lancet*, **2**, 53–56.

DISCUSSION

Pickering: It appears that Dr Murphy has been unable to get pure leucocyte pyrogen, so it isn't possible to be certain about its chemical nature. It has some proteins in it; what about its amino acid constitution?

Murphy [Added in Proof]: We do not propose to look at the amino acid composition of our preparations until we are certain they are homogeneous. We are not sure whether the protein has any small lipid or carbohydrate group attached to it. Any large non-protein component appears to be excluded by a buoyant density of 1·27.

Atkins: I think one of Dr Murphy's strongest arguments is that it is practically pure protein.

Pickering: How does he know that?

Snell: The purified material has just the one band in the conditions that were used to separate the proteins. The method he used to separate the proteins on the gel was the same as he has used to purify the pyrogen. So it is not very strong evidence.

Murphy [Added in proof]: There is much indirect evidence that leucocyte pyrogen is a protein; it is extremely heat-labile, sensitive to trypsin but not to DNAse, RNAse, neuraminidase or a variety of lipid extraction procedures, and it is precipitated by trichloroacetic acid, ammonium sulphate and other reagents commonly used to precipitate or denature proteins (Rafter *et al.*, 1966).

Atkins: The active lipopolysaccharide fraction of bacterial pyrogen has now been obtained essentially free of protein, so the question you raised in your opening remarks, Sir George, as to whether bacterial and endogenous pyrogens are similar can now be partially answered chemically.

Pickering: That wasn't my question. I asked whether leucocyte pyrogen was derived from bacterial pyrogen. If you take the bacterial pyrogen to be lipopolysaccharide, is it that being tagged onto protein, or has leucocyte pyrogen had its molecule altered in some way by a bit of protein being substituted somewhere?

Work: I agree with the chairman that there is no proof yet that there is not a bit of lipopolysaccharide attached to the protein. Dr Murphy hasn't told us whether there is any phosphorus, carbohydrate or lipid in his material. It is obviously a very complex molecule, but until he gets enough to analyse we will not know what it contains besides protein.

Bodel: Using stimuli such as phagocytosis or certain steroids, it is possible to produce leucocyte pyrogen *in vitro* in systems which are apparently free of gram-negative bacterial pyrogen (endotoxin). Therefore, to assume that leucocyte pyrogen contains any endotoxin one would have to postulate that the cells carry with them undetectable endotoxin which can then be transformed into leucocyte pyrogen on appropriate stimulation.

Pickering: I don't think that is likely but ultimately the valid evidence should come from the chemistry of Dr Murphy's product which is not yet pure enough.

Cranston: How did he produce his original white cells?

Murphy [Added in proof]: The white cells were obtained from rabbit peritoneal exudates induced with shellfish glycogen. We know this contains bacterial pyrogen, but we did not bother to remove it by baking for two reasons. First, baking destroys a large part of the ability to attract cells into the peritoneum. Secondly, it seems probable that there is no method of producing an exudate in the peritoneum which is not contaminated with endogenous bacterial pyrogen from the bowel. Our overriding consideration was to get enough material to work with and we are well aware that we have not excluded the possibility that leucocyte pyrogen is derived from bacterial pyrogen. However, if it can be shown that an identical leucocyte pyrogen can be derived from blood leucocytes stimulated by some means other than bacterial pyrogen, then it seems to us that it would be reasonable to suppose that leucocyte pyrogen is not derived from bacterial pyrogen. We are going to perform such an experiment in the near future.

Atkins: Dr Wood recently suggested that the macromolecular component in exudate fluid that activates cells may contain small amounts of endotoxin, because he is unable to distinguish these possibilities with the bioassay he used. I think he can detect the activating effect of about one nanogramme of endotoxin (a subpyrogenic dose when given intravenously) in a system using exudate leucocytes (Moore *et al.*, 1970). On the other hand, it is difficult to extract pyrogen from the unstimulated leucocyte. Since agents that do not contain endotoxin can stimulate leucocytes to release endogenous pyrogen, one would have to postulate that leucocytes contain endotoxin in a non-pyrogenic form and that this agent becomes detectable in a different form as "endogenous pyrogen" after the cell is activated.

Bondy: In the steroid system there has to be either synthesis of leucocyte pyrogen or its release from some precursor stored in the cells. This precursor could be bacterial pyrogen which the cell has picked up before it was exposed to the steroid.

Work: I am sure that there is no complete lipopolysaccharide attached to endogenous pyrogen. There is no definite analytical proof yet that pyrogen is pure protein, and I am not quite sure that there is no degraded lipopolysaccharide there; perhaps a small biologically active group is attached.

Cooper: Rafter and his co-workers (1966) obtained a leucocyte pyrogen which behaved as though it had a molecular weight somewhere between 10 000 and 20 000. This definitely had lipid fragments attached to the molecule.

Murphy [Added in proof]: The suggestion that leucocyte pyrogen contained an essential lipid fraction was subsequently withdrawn. The alkaline sensitivity is due to the oxidation of an essential SH group and can be reversed with mercaptoethanol (Kozak *et al.*, 1968).

Pickering: The real problem here is that there isn't a good source of supply.

Bodel: We have also been trying some simple purification steps, starting with the human blood leucocyte rather than the rabbit exudate leucocyte. Preliminary results indicate that our material focuses below pH 7, whereas rabbit blood leucocyte pyrogen seems to focus above pH 7.

Pickering: So you are saying that there is not a single endogenous pyrogen; are they different when they come from rabbit leucocytes or human leucocytes?

Bodel: Yes, that seems likely. The main pyrogenic activity in preparations of both rabbit and human leucocyte pyrogens appears at about the same molecular weight by Sephadex filtration, but in our experience, a second peak of activity often appears at a higher molecular weight (Bodel, Wechsler and Atkins, 1969). Similar lack of homogeneity occurs with different preparations of leucocyte pyrogen after isoelectric focusing. For example, leucocyte pyrogen obtained from post-phagocytic, frozen-thawed cells (so-called "intracellular pyrogen") appears to be quite different from the material the cell releases into the medium after stimulation. It focuses at pH 6 or below, in contrast to released pyrogen which focuses at nearly pH 7. Also the "intracellular pyrogen" may have a different molecular weight. We are in the process of studying some of these differences. They suggest that the chemical characteristics of leucocyte pyrogen may depend on the cell source as well as the way in which the pyrogenic material is produced.

Work: You would expect species specificity; man and the rabbit might differ in just one or a few amino acid residues and this could lead to different behaviour on isoelectric focusing.

Pickering: When different bacterial pyrogens from different sources are added to leucocytes, do you always get the same substance out of the leucocytes?

Atkins: The products behave similarly in these relatively crude biological tests but these endogenous pyrogens have not as yet been identified in any detail biochemically.

Cranston: Until they are pure you can't tell whether they are the same or not.

Pickering: Can you get an antibody from endogenous pyrogen?

Atkins: Although I believe an antibody has been obtained, it does not appear to be directed against the pyrogenic activity of endogenous pyrogen (Hahn, H. H., and Wood, W. B., unpublished).

Bondy: It has been possible to get antibody against such normal physiological materials of small molecular weight as ACTH (Demura *et al.*, 1966), vasopressin, and cyclic adenylic acid (Steiner *et al.*, 1969). I'm sure some animals somewhere will react to this stuff if it can be got pure enough.

Work: But you have to get enough of it too, which is still a problem.

Landy: You are considering the dose used for immunization, based on the characteristic pyrogen activity. One may be deceived into thinking it is large, in weight units, whereas it may actually be quite inadequate to elicit an immune response. The first requirement for immunogenicity, that of a species of origin difference, is met. This should help to increase the response of the recipient. There are all too many examples in biology where one thinks one is giving a great deal of an active principle, in this case pyrogen, but in weight units it subsequently turns out to be insufficient.

Murphy [Added in proof]: The specific activity of leucocyte pyrogen is such that we can only obtain very small weights of material. From 20 litres of crude leucocyte pyrogen, representing about 200 peritoneal exudates, we would expect to obtain enough apparently pure leucocyte pyrogen to give each of five or ten thousand rabbits a fever of 1°C. The weight of this material would only be 300 μg at most. The minimal dose for an antibody response to a good antigen is about 10 μg. Animals vary greatly in how well they respond to immunization, and of course many doses would be required to get high levels of antibody. Clearly, one would be fortunate to achieve much with such small quantities of material.

Pickering: One way of testing for antigenicity is to see whether the serum from the injected animal will suppress the response. Butler, Harington and Yuill (1940) linked salicylate with casein and produced an immune response. They showed that the serum from the immunized animal would suppress the ordinary effects of salicylates on febrile animals.

Landy: What one wants is not necessarily an antibody that will neutralize but rather one that will react with pyrogen in a specific demonstrable sense.

Cranston: But can you determine that the antibody is reacting with leucocyte pyrogen in a specific sense, unless you can demonstrate that it has an effect on pyrogenicity? Leucocyte pyrogen has not been obtained as a pure product, so an immunoassay may be measuring something other than the pyrogen.

Pickering: You can get bacterial pyrogen pretty pure as a lipopolysaccharide, can't you?

Work: You can get lipopolysaccharides, but very few people have got homogeneous lipopolysaccharides. You have populations of similar molecules; do you call that pure?

Landy: My own view is that it is likely to be physically heterogeneous but the constituents would be of a homogeneous antigenic character.

Pickering: If you extract the lipopolysaccharides from two different gram-negative bacteria are the extracts identical?

Palmer: No, they are not.

Pickering: So there are several molecules which presumably have some sort of chemical configuration in common.

Palmer: If you carry out tests on these different homogeneous lipopolysaccharides the properties differ enormously; some are thermostable and others thermolabile in exactly the same conditions.

Whittet: They react differently to ion-exchange resins too.

Palmer: Yes, one might be taken up by the anionic, another by the cationic resin.

Work: That may be due to differences in the polysaccharide chains, not to biologically active groups which are probably similar in all true lipopolysaccharides. For example, lipopolysaccharide of one group of *Salmonella* can differ by one sugar or acetyl residue from that of another group but they react in the same way in the body, apart from their antigenic activities.

Atkins: By fever assay it can be shown that gram-positive bacteria behave differently from gram-negative bacteria. We have grown a number of gram-positive organisms in culture media that are free of the usual contaminating endotoxins of gram-negative bacteria, and, with the exception of certain strains of staphylococci and streptococci, the filtrates of these organisms are uniformly non-pyrogenic. With *Bacillus subtilis*, diphtheroids and pneumococci, if one injects the whole organism intravenously, heat-killed to avoid the effects of infection, a febrile response appears after a relatively long latent period of one hour (Atkins, 1963). It can be inferred by *in vitro* studies that this response is due to phagocytosis. These data suggest, therefore, that these organisms, unlike gram-negative bacteria, are not shedding pyrogenic lipopolysaccharides into the medium. The soluble pyrogenic agents that appear in growing cultures of staphylococci and streptococci are proteins and apparently act as antigens (like tuberculin) in sensitized hosts and also, in contrast to endotoxin, produce fever after a relatively prolonged latent period (Bodel and Atkins, 1964; Schuh, Hríbalová and Atkins, 1970).

Pickering: Could you get the same effect by injecting inert particles of roughly the same size and number?

Atkins: Berlin and Wood (1964) tried to do this with polystyrene particles and were unable to get fever. I think others have produced fever with agents such as latex but there was evidence that these particles were contaminated with small amounts of endotoxin (Kobayashi and Friedman, 1964). As far as I know Wood hasn't done any *in vitro* studies.

Whittet: It has been done with kaolin (Bennett, 1956) and several other things. I think methylcellulose has been used (Wiedersheim *et al.*, 1953), and some of the dextrans (Bennett, 1952) but of course these too might be contaminated with bacterial pyrogens.

Bangham: Would a reference preparation of either rabbit or human leucocyte pyrogen be practicable at this stage? If it is possible to make a stable one, its use would enable quantitative comparison of results from different laboratories. Provided it could give a parallel transform (e.g. log-) dose response with purified preparations, that's all one needs. If one could produce crude material in some (large) quantity it should be quite straightforward to set up a reference material.

Work: Murphy said that the crude preparations lost activity due to the presence of some proteolytic enzyme. It was when they were pure they were stable.

Bodel: In our experience the purer the material becomes the more unstable it is, losing pyrogenicity rapidly even at 4°C.

Murphy [Added in proof]: Crude leucocyte pyrogen is almost completely stable at 4°C. However, once it has been acidified to pH 3·5 (as part of the filtration process) it loses activity and has a half life of 1–2 days at 4°C. Since further purification makes it stable again, we assume that some proteolytic enzyme is activated by low pH and subsequently remains active. In gel filtration experiments, the purest pyrogen becomes stable, while the less pure material contaminated with a large adjacent protein peak continues to deteriorate. Leucocyte pyrogen is stable in the fractions from isoelectric focusing columns; this is due to the small molecules they contain. When we tried to remove them the pure material promptly became unstable, as Dr Bodel describes. This is why we do not have a satisfactory specific activity for the apparently pure material.

REFERENCES

ATKINS, E. (1963). *Yale J. Biol. Med.*, **35**, 472.

BENNETT, I. L. (1952). *Proc. Soc. exp. Biol. Med.*, **81**, 266.

BENNETT, I. L. (1956). *Bull. Johns Hopkins Hosp.*, **98**, 31.

BERLIN, R. D., and WOOD, W. B. (1964). *J. exp. Med.*, **119**, 715.

BODEL, P. T., and ATKINS, E. (1964). *Yale J. Biol. Med.*, **37**, 130.

BODEL, P. T., WECHSLER, A., and ATKINS, E. (1969). *Yale J. Biol. Med.*, **41**, 376.

BUTLER, G. C., HARINGTON, C. R., and YUILL, M. E, (1940). *Biochem. J.*, **34**, 838–845.

DEMURA, H., WEST, C. D., NUGENT, C. A., NAKAGAWA, K., and TYLER, F. H. (1966). *J. clin. Endocr. Metab.*, **26**, 1297–1302.

KOBAYASHI, G. S., and FRIEDMAN, L. (1964). *Proc. Soc. exp. Biol. Med.*, **116**, 716.

KOZAK, M. S., HAHN, H. H., LENNARZ, W. J., and WOOD, W. B. (1968). *J. exp. Med.*, **127**, 341–357.

MOORE, D. M., CHEUK, S. F., MORTON, J. D., BERLIN, R. D., and WOOD, W. B. (1970). *J. exp. Med.*, **131**, 179.

Rafter, G. W., Cheuk, S. F., Krause, D. W., and Wood, W. B. (1966). *J. exp. Med.*, **123**, 433–444.
Schuh, V., Hríbalová, V., and Atkins, E. (1970). *Yale J. Biol. Med.*, **43**, 31.
Steiner, A. L., Kipnis, D. M., Utiger, R., and Parker, C. (1969). *Proc. natn. Acad. Sci., U.S.A.*, **194**, 367–374.
Wiedersheim, M., Hortlein, W., Huseman, E., and Lotterle, R. (1953). *Arch. exp. Path. Pharmak.*, **217**, 107.

ROLE OF LEUCOCYTES IN FEVER

ELISHA ATKINS* AND PHYLLIS T. BODEL

Department of Experimental Pathology, Scripps Clinic and Research Foundation, La Jolla, California, and Department of Internal Medicine, Yale University School of Medicine, New Haven, Connecticut

DISCOVERY AND SIGNIFICANCE OF ENDOGENOUS PYROGENS IN EXPERIMENTAL AND CLINICAL FEVERS

THE notion that fever and inflammation are intimately linked dates back to the origins of modern pathology. One hundred years ago Billroth (1865) and Weber (1864) independently set out to prove this relationship by injecting pus into animals and observing them for signs of fever. These abortive experiments lay unconfirmed, however, until Menkin's studies, beginning in 1943, culminated in the crystallization of a fever-inducing substance which he extracted from the euglobulin fraction of inflammatory exudates and called "pyrexin" (Menkin, 1956). In retrospect, pyrexin, as well as the material used in the earlier experiments, was almost certainly contaminated with bacterial pyrogens, the complex lipopolysaccharides which form a distinctive part of the cell walls of gram-negative bacteria and are generally known as "endotoxins". These ubiquitous agents, present in air and water, readily pass through bacterial filters and caused many of the so-called "injection fevers" of the early 20th century through their ability to contaminate various biological materials (Bennett and Beeson, 1950).

It was not until 1948 that Beeson showed in experiments in which bacterial pyrogens were carefully excluded that polymorphonuclear leucocytes contain a fever-inducing substance. This brief report was expanded by Bennett and Beeson into two papers (1953*a*, *b*) which remain a landmark in this field, since they provided the first clear proof that animal cells contain a potentially pyrogenic substance and that this agent can be clearly distinguished from the endotoxins of gram-negative bacteria by a number of simple biological criteria: (1) its greater heat lability; (2) its failure to induce a state of unresponsiveness (or pyrogenic tolerance) on repeated intravenous injection; and (3) its ability to produce fevers of equal magnitude in normal recipients and in those rendered tolerant to bacterial pyrogens by daily injections. This lack of so-called "cross-tolerance"

* Supported on sabbatical leave by National Institute of Allergy and Infectious Diseases Special Fellowship 7 F03 AI15698-01A1.

between leucocyte pyrogen and gram-negative bacterial endotoxin, first demonstrated by Bennett and Beeson, has become a standard technique for distinguishing between pyrogens, particularly those of bacterial and tissue origin.

Bennett and Beeson's studies indicated that the only cell from which fever-inducing substances could be extracted was the granulocyte, derived either from exudates or blood or from dermal lesions, such as the Arthus and Shwartzman phenomena, in which this cell predominates. As will be seen, later work has extended the number of cell types that can produce pyrogen (hereafter referred to as "endogenous pyrogen", in contrast to exogenous pyrogens, derived largely from microbes and their products). It has also been since shown that exudate and blood cells differ in the amounts of pyrogen they normally contain and that little can be extracted from blood leucocytes unless they are specifically activated by various stimuli.

Although Bennett and Beeson's work provided evidence for an endogenous source of pyrogen to mediate fever, it did not investigate the steps by which this agent is produced or released during natural or experimentally induced fevers.

The first evidence linking leucocyte pyrogen with fevers induced by endotoxins was presented in a series of studies made by Cranston, Gerbrandy, Goodale, Snell and Wendt in the laboratory of Sir George Pickering. Starting with the observation that intravenous injections of whole blood incubated with endotoxin produced a more rapid onset of fever in human volunteers than did endotoxin alone (Gerbrandy, Cranston and Snell, 1954) these authors went on to demonstrate that the response was critically dependent upon the leucocyte content of the infused blood (Cranston *et al.*, 1956). At about the same time, Atkins and Wood (1955*a*, *b*), using a passive transfer method of Grant and Whalen (1953) as a model, showed that injected endotoxin (typhoid vaccine) was rapidly cleared from the circulation of donor rabbits and that the resulting fever was directly correlated with the appearance and titre of a fast-acting serum pyrogen that had the same properties as leucocyte pyrogen when assayed in normal and pyrogen-tolerant recipient rabbits.

Circulating endogenous pyrogen* has subsequently been shown to be present in fevers induced by a variety of microbes (both gram-negative and gram-positive bacteria and mycobacteria, as well as several viruses and fungi and their products), given intradermally, intravenously or intraperitoneally to several animal species (Atkins and Snell, 1965).

* The terms endogenous pyrogen and leucocyte pyrogen will be used interchangeably in this paper, as evidence to date has implicated with certainty only a few types of leucocytes and tissue macrophages as sources of endogenous pyrogen (see below).

Further, since endogenous pyrogen seems to have a direct action on the thermoregulatory centres, it now seems established that this agent serves as the final common pathway or afferent signal to the central nervous system for the development of fever induced by most, if not all microbes, by immunological reactions to heterologous animal proteins, as well as by a new class of pyrogenic steroids, discussed elsewhere in this symposium (Bondy and Bodel, 1971).

Attempts to demonstrate circulating endogenous pyrogen in human fevers have been less successful, presumably because of the large volume of serum that must be transferred from a febrile to a non-febrile subject to demonstrate this agent (Snell, 1961). However, at least one positive transfer has been reported with autologous plasma taken from a febrile patient after an injection of endotoxin and given back during an afebrile period (Snell *et al.*, 1957). Purulent exudates from patients, however, have been shown to be a potent source of pyrogenic activity when filtered and transfused into afebrile subjects, whereas transudates from non-infectious diseases were regularly non-pyrogenic when passively transferred in the same manner (Snell, 1962).

After this brief review of earlier work demonstrating the significance of endogenous pyrogen in the pathogenesis of various experimental and clinical fevers, we shall devote the rest of this paper to recent evidence bearing on the following questions.

(1) What cells serve as the source of endogenous pyrogen?

(2) What are some of the factors involved in the pathogenesis of fevers in various immune states? Two recent experimental models, using penicillin and the well-known artificial conjugated protein, dinitrophenol-bovine gamma globulin (DNP-BGG), will be primarily discussed in this connexion.

(3) What cellular processes have been identified in the production and release of leucocyte pyrogen?

ROLE OF MONOCYTES AND MACROPHAGES

Using saline extracts of normal tissues, Bennett and Beeson found in their initial studies that the polymorphonuclear leucocyte was the only cell which appeared to contain pyrogen. When a similar survey of normal rabbit tissues (homogenized in the cold and extracted with saline) was undertaken 10 years later by Snell and Atkins, however, significant pyrogenic activity was found in nearly all tissues tested, including heart, lung, liver, spleen, kidney and skeletal muscle (Snell and Atkins, 1965). Relatively large amounts of tissue (1·5–3 g) were required to obtain a single pyrogenic dose but the properties of so-called "tissue pyrogen" were clearly differentiated from gram-negative bacterial endotoxin and resembled those of leucocyte pyrogen. Since tissues of rabbits rendered profoundly granulocytopenic by nitrogen mustard contained nearly as much extractable pyrogen as did those of normal animals, it seemed unlikely that sequestered granulocytes contributed significantly to the observed activity. The suggestion that these tissues may have contained a similar, widely distributed pyrogenic cell, such as the tissue macrophage, seems borne out by subsequent studies demonstrating fever-inducing activity in monocytes and macrophages isolated from several sources.

In the first of these studies to be reported, human blood monocytes obtained from patients with acute leukaemia as well as from normal volunteers were suspended in Krebs–Ringer phosphate buffer with normal serum and were allowed to phagocytize heat-killed staphylococci *in vitro* (Bodel and Atkins, 1967). Large amounts of pyrogen were evolved after incubation for 18 hours. Supernatant fluids from these incubated cells were tested for fever-inducing activity by injection into rabbits. This technique, which avoids the use of human recipients, had been previously shown to be an effective method of assaying human endogenous pyrogen (Bodel and Atkins, 1966). Further studies in rabbits, using either alveolar macrophages (flushed from the lungs by an adaptation of Myrvik's technique) (Atkins, Bodel and Francis, 1967) or peritoneal macrophages (Hahn *et al.*, 1967), clearly associated pyrogen production with this cell or class of cells. Depending on the stimulus used, pyrogen release could also be demonstrated from mononuclear cells of the spleen and mesenteric lymph nodes. When phagocytosis of staphylococci was used as the stimulus, higher ratios of bacteria to cells (20:1 and higher) were required to induce pyrogen release from cell preparations consisting exclusively of mononuclear cells than from blood leucocytes (predominantly granulocytes), in which ratios as low as 5:1 were equally effective (Atkins, Bodel and Francis, 1967). When compared with exudate granulocytes, monocytes from peritoneal exudates continued to release pyrogen for much longer periods (6 to 7 hours) after their initial activation by a substance contained in the exudate fluid (Hahn *et al.*, 1967). In their ability to release large amounts of pyrogen in saline, exudate monocytes resemble exudate granulocytes, which spontaneously release pyrogen in saline. Alveolar macrophages, on the other hand, require a more supportive medium (Krebs–Ringer phosphate buffer and serum) for full activity and are rarely spontaneously active.

Another monocytic cell that has been recently found to be a source of endogenous pyrogen is the Kupffer cell of the liver. Using techniques to separate this component of the reticuloendothelial system from hepatocytes *in vitro*, Dinarello, Bodel and Atkins (1968) have shown that the Kupffer cell can be stimulated to release pyrogen by such standard microbial activators as purified endotoxin, tuberculin and phagocytosis of heat-killed staphylococci, whereas hepatocytes remain inactive.

At present, therefore, it appears that the capacity to produce endogenous pyrogen is shared by a variety of phagocytic cells in both the granulocytic and monocytic series. More critical questions as to what significance can be attached to this association of functions, when these cells acquire this ability in their lifespan and whether eosinophils and basophils can also produce endogenous pyrogen may be answered when techniques permit the separation of large numbers of these cells in a high degree of purity.

POSTULATED MECHANISMS OF FEVER PRODUCTION IN SEVERAL MODELS OF HYPERSENSITIVITY

Hypersensitivity fevers have been demonstrated to a variety of antigens in animals and are clinically well known. In a study designed to determine the cause of clinical fevers resulting from therapy with penicillin, rabbits were immunized with intradermal injections of emulsions of penicillin in adjuvant, intravenous injections of aqueous penicillin, or intramuscular injections of procaine-penicillin (Chusid, 1970).

Fever was produced in sensitized animals by intravenous injections of a penicilloyl–serum protein conjugate prepared by incubating penicillin with serum in alkaline conditions and purified by dialysis against an ion-exchange resin. Aqueous penicillin, on the other hand, was non-pyrogenic over a wide dose range (see Fig. 1). Reactivity to the conjugate could be passively transferred to normal rabbits by serum from sensitized rabbits, as in other models where passive transfer of sensitized serum has produced febrile reactivity to heterologous animal proteins such as bovine serum albumin (Grey, Briggs and Farr, 1961) or human serum albumin (Mott and Wolff, 1966).

In an effort to determine what tissues were the source of endogenous pyrogen in the penicillin model, blood, liver and spleen cells of sensitized animals were incubated with the conjugate in Krebs–Ringer phosphate buffer with serum of normal or sensitized donors and the supernatant fluids assayed *in vivo* for pyrogenic activity. Of these tissues, only the sensitized blood cells released significant amounts of pyrogen *in vitro*, and then only when serum from sensitized rabbits was present, as shown in Fig. 2. In this model, therefore, it appears that antigen–antibody complexes play an essential initiating role in activating these cells to produce pyrogen—an activity that may be correlated with the immediate, profound granulocytopenia that follows intravenous inoculation of the conjugate in sensitized

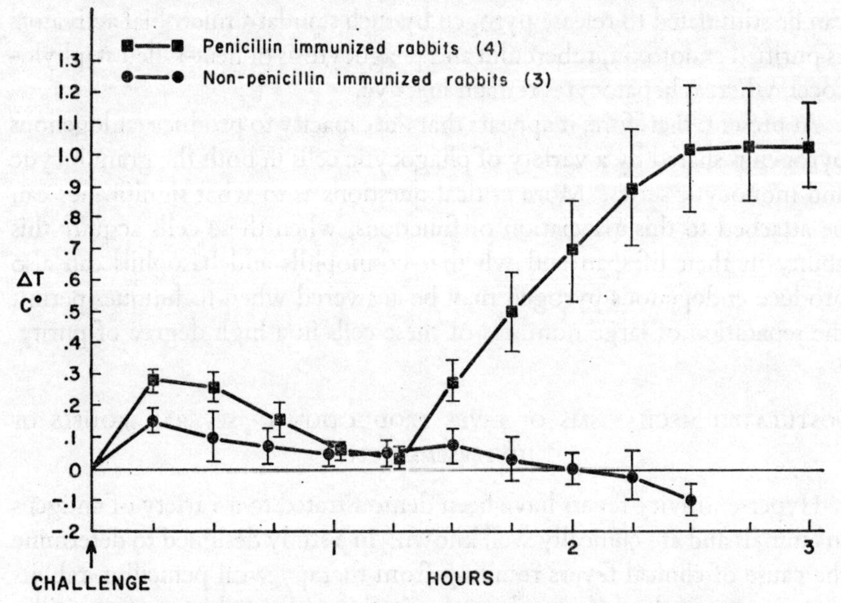

FIG. 1. Mean febrile responses (temperature change in °C ± S.E.M.) of penicillin-immunized and unimmunized rabbits to intravenous challenge with 2 ml of ($1 \cdot 2 \times 10^{-2}$ M-penicilloyl) penicillin–rabbit serum protein conjugate are compared.

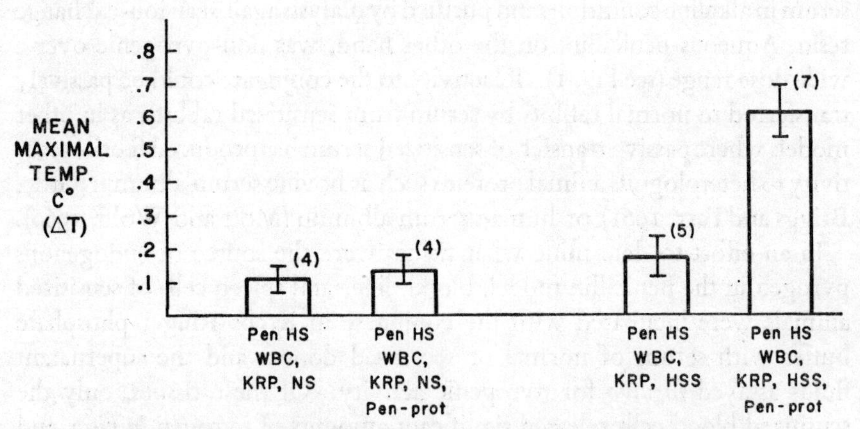

FIG. 2. Mean maximal febrile responses (± S.E.M.) of animals injected intravenously with supernatants of blood cells from rabbits immunized with penicillin (Pen HS WBC). Cells were incubated with Krebs-Ringer phosphate (KRP) and normal rabbit serum (NS) or penicillin hypersensitive serum (HSS) and with and without the penicillin–rabbit serum protein conjugate (Pen–prot). Doses of supernatant injected were equivalent to 1×10^8 white blood cells/dose (10 ml). The numbers in parentheses are the numbers of animals injected.

animals. Root and Wolff (1968) have postulated a similar mechanism in an *in vivo* model employing human serum albumin as the sensitizing antigen. Evidence was obtained that pyrogenic antigen–antibody complexes appear in the circulation of sensitized rabbits shortly after intravenous injection of this antigen. Furthermore, characteristic hypersensitivity fevers were produced in unsensitized animals given soluble antigen–antibody complexes prepared *in vitro* in the zone of antigen excess.

In a model designed to distinguish the pyrogenic effects of delayed hypersensitivity from those of immediate hypersensitivity mediated by humoral

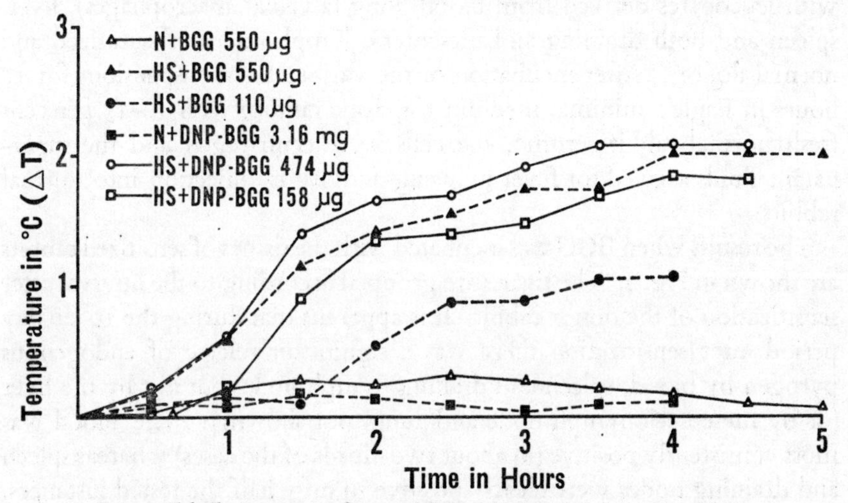

FIG. 3. Febrile responses of normal (N) and DNP-BGG sensitized (HS) rabbits to various doses of bovine gamma globulin (BGG) and dinitrophenol–bovine gamma globulin (DNP-BGG). Each curve represents the average of 2 recipients, challenged once between days 14 and 17 after sensitization.

antibody, Dr J. D. Feldman and one of us (E.A.) have recently used dinitrophenol conjugated with bovine gamma globulin (DNP–BGG) as the sensitizing antigen. Rabbits were immunized with a total of 10 mg DNP-BGG incorporated in complete Freund's adjuvant, the injections being made into the foot pads and four subcutaneous sites at the shoulders and haunches. This technique induces a state of delayed hypersensitivity to the carrier protein (BGG) from about the 5th to 8th days, during which antibodies are not detectable by the Ouchterlony technique. On the other hand, precipitating antibodies to the conjugate are present throughout this period.

Sensitized animals develop marked febrile responses to microgramme amounts (550 and 474 µg) of both carrier and conjugate given intravenously, as indicated in Fig. 3. The responses to the two agents are similar in appearance and have their onset after the usual latent period of 45–60 minutes characteristic of most allergic fevers. With smaller doses (110 and 158 μg), the latency is prolonged and the fever reduced, as is shown in Fig. 3. Unsensitized animals have little or no febrile response to the same or greater doses, indicating the specificity of this feature of hypersensitivity and the lack of pyrogens in the reagents.

In an attempt to determine which tissues were responsible for releasing the endogenous pyrogen that mediates this form of fever, both BGG (the carrier) and DNP-BGG (the imunizing agent) were incubated overnight with leucocytes derived from blood, lung (alveolar macrophages), liver, spleen and both draining and mesenteric lymph nodes of sensitized and normal donors. After incubation of the various tissue suspensions for 18 hours in Eagle's minimal medium for tissue culture with 10–15 per cent fresh normal rabbit serum*, the cells were centrifuged and the supernatant fluids assayed for fever inducing-activity by injection into normal rabbits.

The results when BGG was incubated with the tissues of sensitized rabbits are shown in Fig. 4. The tissues are grouped according to the interval after sensitization of the donor rabbit. It is apparent that during the 10–20 day period after sensitization there was a significant release of endogenous pyrogen by blood, spleen and draining lymph node, but not by the liver (or by mesenteric lymph node and lung, not shown)†. The blood was most consistently positive (in about two-thirds of the cases) whereas spleen and draining nodes were clearly positive in only half the tested instances. Thereafter, 21–35 days after sensitization, as shown by the bars at the right of Fig. 4, there was only an occasional positive response and none from the draining lymph node after day 14. Since circulating antibodies to BGG can be assumed to be still present at 3–5 weeks, the difference between these *in vitro* results at the early and late intervals suggests that the reactive agent is present only early after sensitization (perhaps gamma M[IgM]) and disappears with time. Alternatively, the response may not depend on humoral antibody, but may involve the generation of an intermediate substance (by the action of specific antigen on sensitized lymphocytes) which, in turn, stimulates leucocytes to produce endogenous pyrogen. One might classify this material as a lymphokine, to which analogous substances such as macrophage-inhibiting factor belong (Dumonde *et al.*, 1969).

* The centrifuged blood cells (with some 10-15 ml overlying autologous plasma) were resuspended in Eagle's minimal medium lacking Ca^{++}.

† Since we were unable to activate the lung macrophages of these rabbits by such regularly effective agents as phagocytosis (of staphylococci), tuberculin or purified endotoxin, as well as with BGG and DNP-BGG, despite using various media and conditions, we prefer not to draw any conclusions about these cells in this system.

To determine whether humoral antigen–antibody complexes might be responsible for the *in vitro* activation of these sensitized tissues, perhaps by a mechanism similar to phagocytosis, blood cells of both normal and sensitized rabbits were incubated in several different media overnight with various dosages of antigen–antibody complexes of DNP–BGG and anti-DNP–BGG, as well as BGG and anti-BGG, both prepared by adding antigen to antisera *in vitro*. Precipitates and soluble complexes in the zone

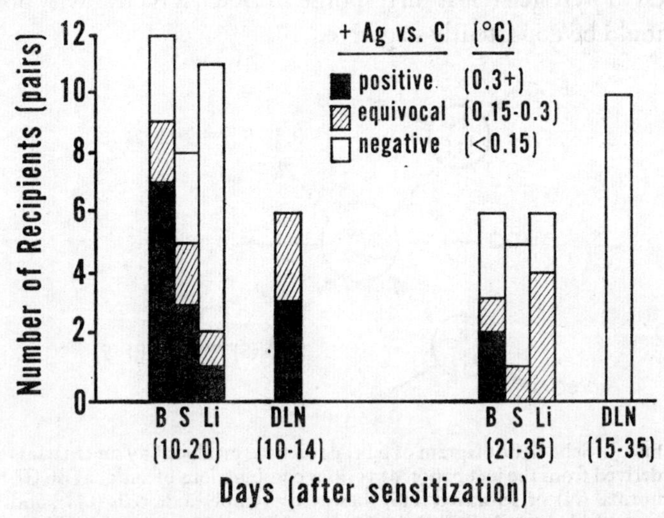

FIG. 4. Relative reactivity of various tissues (from rabbits sensitized to DNP–BGG) incubated *in vitro* with BGG. Average of two febrile responses to supernatants of tissues incubated either with or without antigen (+Ag *vs.* C) are compared for each experiment and differences recorded in the three categories given in the key. Numbers of differences thus derived ("recipient pairs") for each tissue are plotted against days after sensitization of donor rabbits. Tissues used: blood: B; spleen: S; liver: Li; draining lymph nodes: DLN.

of antigen excess were formed. In no instance was an unequivocal release of endogenous pyrogen obtained by such mixtures, despite dosages of antigen up to ten to forty-fold higher than those that produced fever when injected intravenously. Since an equivalent dose of antigen caused significant release of endogenous pyrogen from sensitized blood cells, it is apparent that conventional antigen–antibody complexes are not primarily responsible for the activation of these sensitized cells by specific antigen.

The failure to activate tissues by antigen-antibody complexes *in vitro* caused us to consider the possibility that release of endogenous pyrogen *in vitro* was due to factors operative in delayed hypersensitivity and that therefore the sensitized lymphocyte might play an important intermediate role. A glance at the tissues that were found to be reactive to BGG *in vitro* shows them to be composed of both lymphocytes and pyrogen-producing ("effector") cells. Significant numbers of lymphocytes are present in the blood, spleen and draining lymph nodes, along with granulocytes (blood) or macrophages (spleen and draining nodes). The two inactive tissues, liver and mesenteric nodes, on the other hand, are characterized by the presence of predominantly effector cells (Kupffer cells) or lymphocytes, respectively. If the lymphocyte were a non-reactive cell but an essential intermediate in the process of pyrogen release in response to BGG, it is clear why both these tissues should be consistently negative.

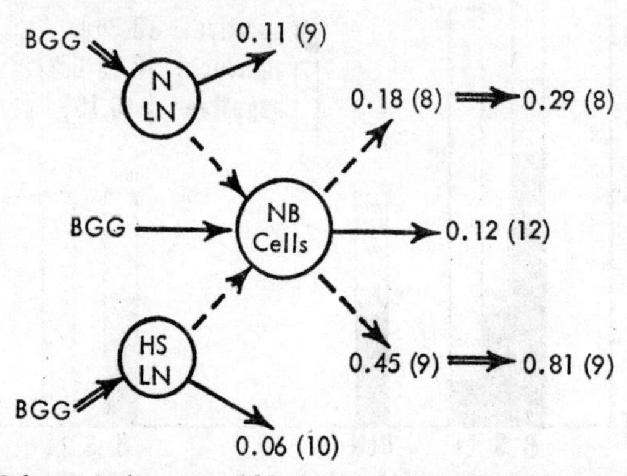

FIG. 5. Schematic diagram of febrile reactions produced by supernatants derived from the incubation of various combinations of antigen (BGG), normal (N) or sensitized (HS) mesenteric lymph node cells (LN), and normal blood cells (NB Cells) *in vitro*. Values at each arrow indicate average fevers (°C) for each combination of normal or sensitized cells shown in the upper and lower halves of the figure, respectively. Numbers of recipients are shown in parentheses.

To test the hypothesis that lymphocytes play an intermediate role in activating other effector cells in this model of hypersensitivity, aliquots of normal blood leucocytes in dosages of about 1×10^8 cells were incubated with lymphocytes from either normal or sensitized mesenteric lymph nodes, together with BGG, for 18 hours. As controls, blood cells were incubated with either normal or sensitized lymph node cells alone or with BGG alone. The supernatants from these cultures were then injected intravenously into rabbits to determine the pyrogenic response. The average results of a number of such experiments, using five donor rabbits 14–20 days after sensitization with DNP-BGG, are presented in Fig. 5, which also shows the results of incubating the antigen with either normal or sensitized lymph node cells alone. Several findings are evident:

(1) Incubation of BGG with either normal or sensitized lymph node cells does not produce detectable amounts of endogenous pyrogen.

(2) Incubation of normal blood cells with sensitized lymphocytes alone

produces a small but somewhat greater average release of endogenous pyrogen than does incubation with normal lymphocytes alone (average fever, 0·45°C compared to 0·18°C).

(3) Addition of the carrier antigen, BGG, significantly increases the production of endogenous pyrogen from blood cells incubated with sensitized lymphocytes, but not from blood cells incubated with normal lymphocytes (average fever, 0·81°C compared to 0·29°C).

(4) Normal blood cells incubated alone with BGG release little or no endogenous pyrogen.

Taken together, these data are consistent with the hypothesis that BGG reacts with specifically sensitized lymphocytes to release a non-pyrogenic intermediate substance or "lymphokine" that activates unsensitized blood cells to produce pyrogen *in vitro*.

Critical evidence to support this hypothesis would obviously be the isolation of a soluble factor produced by the reaction of sensitized lymphocytes and antigen and capable of activating normal cells. Efforts are now under way to determine if such a factor can be detected, utilizing a more nearly pure source of lymphocytes, namely the thoracic duct. Although crude preparations of macrophage-inhibiting factor [produced by overnight incubation of PPD (purified protein derivative) with specifically sensitized guinea pig lymph node cells in Eagle's minimal medium] do activate normal rabbit blood cells, the results are clouded by the fact that the PPD employed, though non-pyrogenic when given intravenously in the dosages used, activates normal rabbit blood cells to some extent when incubated alone with them.

PRODUCTION AND RELEASE OF PYROGEN BY EFFECTOR CELLS

In the final section, we shall discuss briefly some of the cellular processes involved in the production and release of endogenous pyrogen. A number of investigators have studied the mechanism by which leucocytes became activated to produce and release endogenous pyrogen. Initial work focused primarily on the mechanism of the release of pyrogen from polymorphs derived from acute peritoneal exudates of rabbits. In a series of papers, Wood and his co-workers have demonstrated that when such cells are suspended in saline release of pyrogen begins almost at once and continues for 1–2 hours. If the cells are suspended in serum or plasma, or if K^+ is added to the saline, release of pyrogen is suppressed. Pyrogen release is not inhibited when puromycin is present but it can be blocked by sulphydryl-reactive enzyme inhibitors, including iodoacetate and arsenite. There is no release during incubation at 4°C. Only small amounts of intracellular endogenous pyrogen can be detected in these cells at any time (Kaiser and Wood, 1962*a*, *b*; Berlin and Wood, 1964*a*; Hahn *et al.*, 1970*a*). From these observations it appears that exudate cells are already "activated" at the time of harvest, perhaps by a macromolecular substance with activating properties in exudate fluid (Moore *et al.*, 1970). For pyrogen to be produced in these cells, then, some metabolic activity but not protein synthesis is required. Presumably, a non-pyrogenic precursor is present in the cell at the time of harvest which then is rapidly converted to pyrogen and released under appropriate conditions of incubation.

To examine early steps in cell activation, a number of studies have recently been made with an inactive cell—the blood leucocyte of rabbits or man (Berlin and Wood, 1964*b*; Bodel, Dillard and Bondy, 1968; Moore *et al.*, 1970; Nordlund, Root and Wolff, 1970; Bodel, 1971). Some preliminary information is also available about the mechanisms of pyrogen production by mononuclear cells from human blood and rabbit tissues (Bodel and Atkins, 1967; Atkins, Bodel and Francis, 1967). None of these cell types contain detectable pyrogen initially, and they do not release pyrogen spontaneously, whether incubated in saline or in the usual serum–buffer media. When stimulated by agents such as endotoxin, phagocytosis or microbial antigens, a series of steps is initiated which results in release of pyrogen. We shall present the results of studies investigating this process in a system of human blood leucocytes stimulated by phagocytosis of heat-killed staphylococci as illustrative of the process, although only a few of these results have been confirmed in systems using other cell tissues and stimuli.

When leucocytes are incubated with heat-killed staphylococci pyrogen begins to be detectable in the medium after about two hours. At the same time small amounts of intracellular pyrogen can be detected in extracts of washed cells. The pyrogen content of the medium increases rapidly for the next four hours (from two to six hours after addition of staphylococci), but then increases only very slowly for the remainder of the standard 18-hour incubation period. During the time of rapid pyrogen release, intracellular concentrations of pyrogen remain very low. After six hours, however, intracellular concentrations of pyrogen rise, coincidentally with the slower release of pyrogen into the medium.

Several distinct steps leading to the production and release of pyrogen can be delineated in this system. These are summarized diagrammatically in Figs. 6 and 7.

(1) *Cell activation*

Initial interaction of a stimulating agent with the cell is required for activation. This usually occurs rapidly, and in almost all systems studied

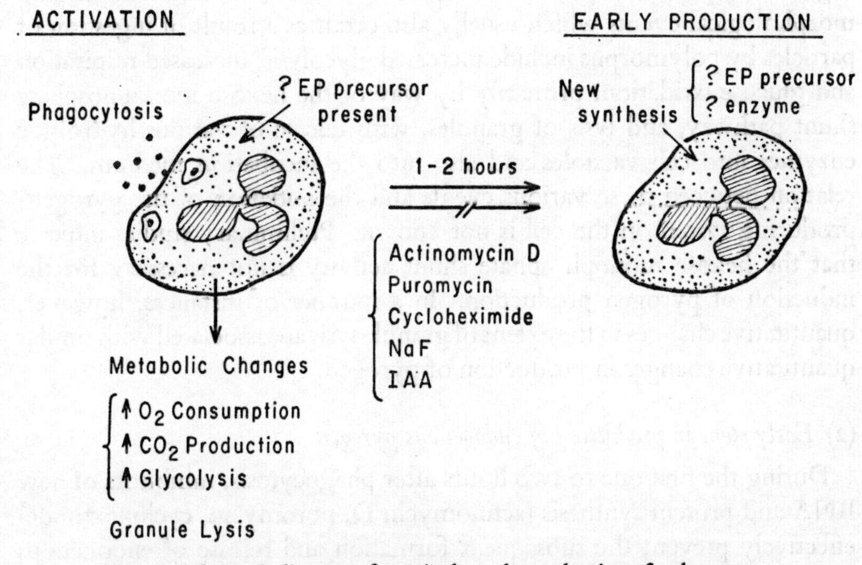

FIG. 6. Schematic diagram of steps in the early production of endogenous pyrogen (EP) by human blood leucocytes stimulated by phagocytosis of heat-killed staphylococci. NaF: sodium fluoride; IAA: iodoacetate.

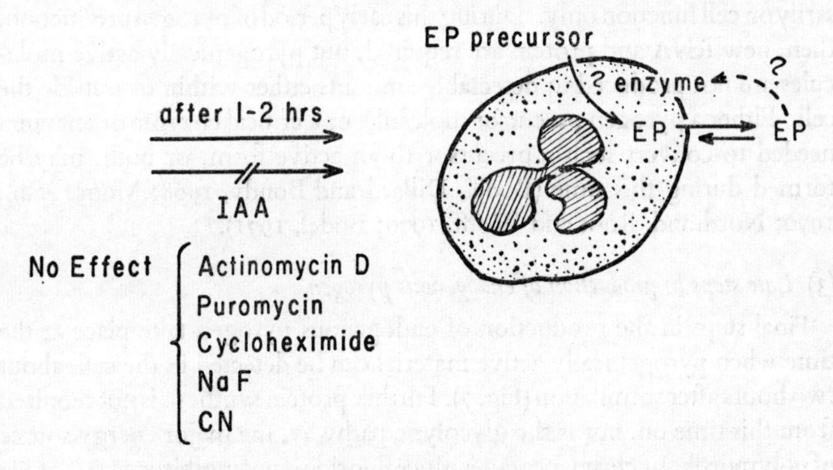

FIG. 7. See Fig. 6. Steps in late production and release of endogenous pyrogen (EP). IAA: iodoacetate; NaF: sodium fluoride; CN: cyanide.

requires the presence of serum or plasma. In phagocytosis, ingestion of the organism is presumably the inciting event. A number of metabolic and morphological events which usually also occur as a result of ingestion of particles by polymorphs include increased glycolysis, increased respiration and glucose oxidation, primarily by way of the hexose monophosphate shunt pathway, and lysis of granules, with release of various hydrolytic enzymes into cell vacuoles and also into the incubation medium. The relation between these various events and the initiation of the pyrogen-producing system in the cell is not known. Preliminary studies indicate that the hexose monophosphate shunt activity is not necessary for the induction of pyrogen production. In a number of instances, however, quantitative changes in the extent of granule lysis are associated with similar quantitative changes in production of pyrogen.

(2) *Early steps in production of endogenous pyrogen*

During the first one to two hours after phagocytosis, inhibitors of new RNA and protein synthesis (actinomycin D, puromycin, cycloheximide) effectively prevent the subsequent formation and release of endogenous pyrogen. Neither intracellular nor extracellular pyrogen develops. Sodium fluoride also inhibits pyrogen production when it is added one hour after the bacteria, but, like actinomycin and puromycin, it has no effect when added at two hours. Iodoacetate is partially inhibitory when added at either time, but because high concentrations of this inhibitor alter the pyrogenic properties of endogenous pyrogen, the effects of this inhibitor are not necessarily on cell function only. During this early period of pyrogen production, then, new RNA and protein are required, but pyrogenically active molecules are not produced in detectable amounts either within or outside the cell. Either a pyrogen precursor molecule, or a critical enzyme or enzymes needed to convert such a precursor to an active form, or both, may be formed during this time (Bodel, Dillard and Bondy, 1968; Moore *et al.*, 1970; Nordlund, Root and Wolff, 1970; Bodel, 1971).

(3) *Late steps in production of endogenous pyrogen*

Final steps in the production of endogenous pyrogen take place at the time when pyrogenically active material can be detected in the cell, about two hours after stimulation (Fig. 7). Further protein synthesis is not required from this time on, nor is the glycolytic pathway, the major energy source of polymorphonuclear leucocytes, since fluoride is not inhibitory*. Cyanide

* Hahn and co-workers (1970*a*) have presented evidence from studies of exudate cells pretreated with sodium fluoride that glycolytic energy is required for the release of endogenous pyrogen.

is also without effect. Intact cell structure, however, appears to be essential, since numerous attempts to induce the formation of pyrogen in activated cell homogenates have been unsuccessful (Hahn *et al.*, 1970*b*; Bodel, 1971).

(4) *Release of pyrogen*

Release of endogenous pyrogen, as indicated above, appears not to require energy. It occurs from two to six hours after phagocytosis, as soon as pyrogen is produced intracellularly. It has been observed by us and others (Root, Nordlund and Wolff, 1970) that larger amounts of endogenous pyrogen are produced if the incubation volume is increased. Further investigation of this finding indicates that the concentration of some released product, possibly endogenous pyrogen itself, becomes a limiting factor in preventing further release, and, in turn, production. As endogenous pyrogen accumulates in the medium, release slows, but a new burst of production and release can be induced by substituting medium which does not contain pyrogen. This new burst of production and release is not dependent on new protein synthesis. This equilibrium may perhaps represent control of the activity of a converting and/or releasing enzyme by a reaction product. In spite of many attempts, no convincing evidence has yet been obtained for the existence of such an enzyme or enzymes (Bodel, 1971). Neither is there evidence to suggest that new synthesis of the endogenous pyrogen molecule itself results from cell activation. In studies in which cells were incubated with [^{3}H]phenylalanine before phagocytosis, no increased labelling of protein in a fraction containing pyrogen was observed (Bodel, 1971). Even if a pyrogen precursor is newly synthesized during this early period, it seems likely that a simple, perhaps enzymic, step or series of steps is also required during the later period to convert this precursor to an active pyrogen, since pyrogen does not become detectable intracellularly until two hours after activation of the cell.

UNANSWERED QUESTIONS

We should like to close this brief review of the role of leucocytes in the development of fever by noting some questions which remain to be answered, both in the broad area of the physiology of endogenous pyrogen as well as in the areas selected for discussion here.

(1) What cell type or types are involved in the pathogenesis of variously induced fevers? Can tumour cells produce pyrogen or are immunological factors involved? Can non-pyrogenic intermediates be implicated in

hypersensitivity and other fever models? What part, if any, does complement play in this process? Does the cell type involved in turn determine the clinical characteristics of the fever?

(2) Once cells become activated, what biochemical steps occur in the production and release of endogenous pyrogen? Is this molecule present in a non-pyrogenic precursor form in many cells, or is it newly synthesized? What is the signal for production? Is a process of induction or repression of a genetically controlled enzyme a likely explanation? What factors control its release?

(3) Does endogenous pyrogen act at sites other than the hypothalamus, perhaps as a stimulus to further inflammation? Does it induce release of endogenous pyrogen from non-active cells? Such a finding may explain the usual biphasic febrile response to a sufficient dosage of pyrogen given intravenously—whether exogenous or endogenous.

(4) Finally, what is the fate of endogenous pyrogen in the circulation? Is it removed by the reticuloendothelial system, excreted, or metabolized to inactive products?

Although fever has been known as a sign of illness since the foundations of medicine were laid by the extraordinary observations of Hippocrates on febrile patients, it has only been in the last 20 years or so, since the definitive discovery of leucocyte pyrogen by Beeson, that we have gained a significant understanding of some of the mechanisms involved in this disease process. However, there is still much that is unknown about the nature of the many different events that produce fever. Nor, at a more philosophical level, do we have a clear idea of what advantages fever provides for the host. It seems unlikely that this universal response of warm-blooded animals would have survived in an evolutionary sense if it did not confer an essential defence against disease.

SUMMARY

A brief review is given of the discovery of leucocyte pyrogen and the role of circulating endogenous pyrogen in various experimental models of fever.

More recent work dealing with the isolation of endogenous pyrogen from cells other than granulocytes is summarized. Sources discussed include pulmonary alveolar macrophages, blood monocytes, spleen and lymph node mononuclear cells and Kupffer cells of the liver.

The probable series of events leading to the development of fever is discussed in two experimental models of hypersensitivity in which rabbits have been sensitized to the following agents: (1) penicillin; and (2) DNP-BGG (dinitrophenol conjugated to bovine gamma globulin).

Finally, various cellular events involved in the production and release of endogenous pyrogen from leucocytes are discussed in a system in which human peripheral blood leucocytes have been activated by phagocytosis of heat-killed staphylococci.

Acknowledgements

This is publication number 424 from the Department of Experimental Pathology, Scripps Clinic and Research Foundation, La Jolla, California. The work was supported by United States Public Health Service Grants AI-07007 and AI-01564.

The authors are deeply indebted to Dr Joseph D. Feldman in whose laboratory one of us (E.A.) has just concluded his sabbatical year, 1969–1970. Dr Feldman has generously contributed his invaluable knowledge, time and funds to one of the projects discussed here as well as reviewing the entire manuscript.

REFERENCES

Atkins, E., Bodel, P., and Francis, L. (1967). *J. exp. Med*, **126,** 357–384.

Atkins, E., and Snell, E. S. (1965): in *The Inflammatory Process*, pp. 495–534, ed. Zweifach, E. W., Grant, L., and McCluskey, R. T. New York: Academic Press.

Atkins, E., and Wood, W. B., Jr (1955*a*). *J. exp. Med.*, **101,** 519–528.

Atkins, E., and Wood, W. B., Jr (1955*b*). *J. exp. Med.*, **102,** 499–516.

Beeson, P. B. (1948). *J. clin. Invest.*, **27,** 524.

Bennett, I. L., Jr, and Beeson, P. B. (1950). *Medicine*, **29,** 365–400.

Bennett, I. L., Jr, and Beeson, P. B. (1953*a*). *J. exp. Med.*, **98,** 477–492.

Bennett, I. L., Jr, and Beeson, P. B. (1953*b*). *J. exp. Med.*, **98,** 493–508.

Berlin, R. D., and Wood, W. B., Jr (1964*a*). *J. exp. Med.*, **119,** 697–714.

Berlin, R. D., and Wood, W. B., Jr (1964*b*). *J. exp. Med.*, **119,** 715–726.

Billroth, T. (1865). *Langenbecks Arch. klin. Chir.*, **6,** 372.

Bodel, P. (1971). *Yale J. Biol. Med.*, in press.

Bodel, P., and Atkins, E. (1966). *Proc. Soc. Biol. Med.*, **121,** 943–946.

Bodel, P., and Atkins, E. (1967). *New Engl. J. Med.*, **276,** 1002–1008.

Bodel, P. T., Dillard, M., and Bondy, P. K. (1968). *Ann. Int. Med.*, **69,** 875–879.

Bondy, P. K., and Bodel, P. T. (1971). This volume, pp. 101–110.

Chusid, M. J. (1970). *Penicillin Allergy and Hypersensitivity Fever.* Thesis, Yale University School of Medicine.

Cranston, W. I., Goodale, F., Jr, Snell, E. S., and Wendt, F. (1956). *Clin. Sci.*, **15,** 219–226.

Dinarello, C. A., Bodel, P., and Atkins, E. (1968). *Trans. Ass. Am. Physns*, **81,** 334–344.

Dumonde, D. C., Wolstencroft, R. A., Panayi, G. S., Matthew, M., Morley, J., and Howson, W. T. (1969). *Nature, Lond.*, **224,** 38–44.

Gerbrandy, J., Cranston, W. I., and Snell, E. S. (1954). *Clin. Sci.*, **13,** 453–459.

Grant, R., and Whalen, W. J. (1953). *Am. J. Physiol.*, **173,** 47–54.

Grey, H. M., Briggs, W., and Farr, R. S. (1961). *J. clin. Invest.*, **40,** 703–706.

Hahn, H. H., Char, D. C., Postel, W. B., and Wood, W. B., Jr (1967). *J. exp. Med.*, **126,** 385–394.

HAHN, H. H., CHEUK, S. F., MOORE, D. M., and WOOD, W. B., JR (1970*a*). *J. exp. Med.*, **131**, 165–178.
HAHN, H. H., CHEUK, S. F., ELFENBEIN, C. D. S., and WOOD, W. B., JR (1970*b*). *J. exp. Med.*, **131**, 701–710.
KAISER, H. K., and WOOD, W. B., JR (1962*a*). *J. exp. Med.*, **115**, 27–36.
KAISER, H. K., and WOOD, W. B., JR (1962*b*). *J. exp. Med.*, **115**, 37–47.
MENKIN, V. (1956). *Biochemical Mechanisms in Inflammation*, 2nd ed. Springfield, Illinois: Thomas.
MOORE, D. M., CHEUK, S. F., MORTON, J. D., BERLIN, R. D., and WOOD, W. B., JR (1970). *J. exp. Med.*, **131**, 179–188.
MOTT, P. D., and WOLFF, S. M. (1966). *J. clin. Invest.*, **45**, 372–379.
NORDLUND, J. J., ROOT, R. K., and WOLFF, S. M. (1970). *J. exp. Med.*, **131**, 727–743.
ROOT, R. K., NORDLUND, J. J., and WOLFF, S. M. (1970). *J. Lab. clin. Med.*, **75**, 679–693.
ROOT, R. K., and WOLFF, S. M. (1968). *J. exp. Med.*, **128**, 309–323.
SNELL, E. S. (1961). *Clin. Sci.*, **21**, 115–124.
SNELL, E. S. (1962). *Clin. Sci.*, **23**, 141–150.
SNELL, E. S., and ATKINS, E. (1965). *J. exp. Med.*, **121**, 1019–1038.
SNELL, E. S., GOODALE, F., JR, WENDT, F., and CRANSTON, W. I. (1957). *Clin. Sci.*, **16**, 615–626.
WEBER, O. (1864). *Dt. klin. ther. Wschr.*, **16**, 461–493.

DISCUSSION

Pickering: Dr Atkins, you told us that when white cells are phagocytosing killed cocci of various kinds, in the early stages inhibitors of protein synthesis will inhibit the release of pyrogen and in the later stages they won't. Is this peculiar to phagocytosis or do bacterial pyrogens in general behave like this, and will the antigen-antibody reaction also have the same system of inhibitors?

Bodel: I have only studied the system of phagocytosis in detail. Although we don't know the time-sequence of inhibition, we do know that puromycin will also inhibit pyrogen release when cells are stimulated by pyrogenic steroids and old tuberculin, and others have used endotoxin (Moore *et al.*, 1970). My guess is that the steps of production and release of pyrogen will turn out to be very similar regardless of the stimulus used.

Pickering: If you make blobs of histamine and antigen on the skin of a pollen-sensitive patient and prick through both blobs, the difference in time between the appearance of the two weals is only a few seconds. If one followed the analogy, one would think that the release of pyrogen from your white cells, Dr Atkins, in an antigen-antibody reaction might be due to something on the surface which was releasing something preformed. But if the inhibitors work in this way presumably it is something quite different.

Atkins: In models using phagocytosis, Dr Bodel has shown that there is no detectable pyrogenic activity either intracellularly or extracellularly for about two hours, although phagocytosis occurs immediately.

Bodel: This is true even when quite large numbers of cells are used, for example 1×10^6 cells per dose.

Cranston: There is a difference between the granulocyte stimulated by phagocytosis and the exudate cell; the release of leucocyte pyrogen from granulocytes harvested from a peritoneal exudate depends on the concentration of potassium in the medium (Berlin and Wood, 1964). The release of pyrogen from a leucocyte stimulated by phagocytosis is independent of the external electrolyte concentration. This cell will go on releasing pyrogen regardless of the external potassium concentration.

Atkins: This explains the situation *in vivo* which has always been puzzling. *In vivo* there are physiological concentrations of potassium and obviously there is fever. The exudate cell is a very special case.

Landy: Dr Atkins, at present immunologists are much impressed with the potential importance of mediators of cellular immunity (Lawrence and Landy, 1970). Were they to be exposed to your findings they might be inclined to attribute the production of endogenous pyrogen to the lymphocyte as much as to the macrophage. As I see it, you believe that the macrophage plays the key role in causing the sensitive lymphocyte to respond efficiently to the antigen. There is now a considerable literature on the mediators of delayed-type hypersensitivity elaborated by the lymphocyte, and their profound effect on the macrophage (Lawrence and Landy, 1970). Your problem is how to sort out the producing cell from the responsive cell, and most of all to avoid confusing cause with effect. By extrapolating from immunological studies the macrophage is now generally ascribed a *helper* function in the efficient presentation of antigen to the specifically sensitive cell (lymphocyte). This being so I would look for the production of endogenous pyrogen, as another mediator produced by these cells, in the now widely studied established continuous cell-lines derived from human peripheral blood lymphocytes. If my points are valid you should find endogenous pyrogen free in the culture medium of such cell-lines.

Atkins: These cell-lines would have to contain effector cells as well as lymphocytes since, from what we know at present, lymphocytes alone will not produce endogenous pyrogen.

Landy: This approach might be a way to check that.

Bodel: Would phytohaemagglutinin-stimulated cells be appropriate ones to study?

Landy: Generally such activated cells make the recognized mediators.

Bodel: We have tested peripheral blood cells five days after stimulation with phytohaemagglutinin and were unable to detect pyrogen production.

Landy: That would perhaps make it less certain, but I still think it would be worth some careful repeated experiments with phytohaemagglutinin.

And I would, nonetheless, examine the products of the established human lymphocytic cell-lines.

Atkins: The lymphocyte does not seem to liberate any pyrogen under any of the conditions in which we incubated it with the antigen to which it is sensitive. So one has to assume an intermediate that works on an effector cell population which may be granulocyte, macrophage, or any other.

Landy: What you say also applies broadly to delayed hypersensitivity. The difference between responses to phytohaemagglutinin and conventional antigens is that the former do *not* require the participation of macrophages whereas the latter do. However, it is believed that here too these cells have a helper function, since if the lymphocytes are subjected to test as a suspension of pellet, the role of the helper cells (macrophages) is correspondingly reduced.

REFERENCES

BERLIN, R. D., and WOOD, W. B. (1964). *J. exp. Med.*, **119**, 697.

LAWRENCE, H. S., and LANDY, M. (eds) (1970). *Mediators of Cellular Immunity*. New York: Academic Press.

MOORE, D. M., CHEUK, S. F., MORTON, J. D., BERLIN, R. D., and WOOD, W. B. (1970). *J. exp. Med.*, **131**, 179.

MECHANISM OF ACTION OF PYROGENIC AND ANTIPYRETIC STEROIDS *IN VITRO*

PHILIP K. BONDY and PHYLLIS BODEL

Department of Internal Medicine, Yale University School of Medicine, New Haven, Connecticut

THE demonstration by Kappas and co-workers (1957) that certain steroids could produce fever when injected into human subjects offered an unusual opportunity to investigate specific pathways involved in the production of fever. The steroids are much simpler molecules than the bacterial or particulate materials commonly used as stimuli in studies of the pathogenesis of fever. Their use in an *in vitro* model, therefore, would result in the introduction of a simple chemical into the pyrogen-releasing system, and might lead to an understanding of some basic mechanisms involved in this process. We shall present data from such studies in an *in vitro* system using human blood leucocytes. Although we are far from achieving a definitive answer, the system seems useful.

The characteristics of the steroid pyrogens, as determined by injection studies in man, are quite specific (Kappas and Palmer, 1963). The steroid must have a 5β hydrogen, since 5α steroids are inactive; a 3α-hydroxyl must also be present. If the hydroxyl group is oxidized to a ketone pyrogenic activity is reduced or lost, and the hydroxyl group must not be blocked by esterification. Certain other modifications of the molecule alter the pyrogenicity of the substance (Table I).

TABLE I

CHEMICAL CHARACTERISTICS OF STEROIDS WHICH ARE PYROGENIC *in vivo*

Configurations required:

- 5β configuration (A:B *cis*)
- 3α-hydroxyl

Configurations which do not affect activity:

- 11-ketone
- 20-ketone
- 20-hydroxyl

Configurations which reduce pyrogenic activity:

- 3-ketone
- 11β-hydroxyl

Configurations which eliminate activity:

- 17α-hydroxyl conjugation on 3-hydroxyl

When steroid pyrogens are injected intramuscularly into human subjects there is a latent period of four to eight hours before the fever begins. If the steroid is given by rapid intravenous injection it is rapidly removed from the blood and fever does not develop. Thus it appears that persistence of steroid during the latent period is required, perhaps in a tissue site, for the steroid to be pyrogenic. The steroid pyrogens are specific for human beings; no other animal studied so far responds to them by developing fever (Kappas and Palmer, 1963). In contrast, the bacterial pyrogen endotoxin produces an effect in less than an hour, is extremely effective when given by rapid intravenous injection, and affects a large variety of experimental animals as well as human beings.

When a variety of cell types are stimulated by agents such as endotoxin or by phagocytosis, a small protein, "endogenous pyrogen", is released. Bodel and Atkins (1966) have shown that endogenous pyrogen released from human cells *in vitro* will cause characteristic fever in rabbits. Thus it was of interest to investigate the action of steroid pyrogens in an *in vitro* system to determine whether they would also cause release of endogenous pyrogen from human cells. The methods used are briefly as follows: A leucocyte-rich fraction is separated from human heparinized blood by dextran sedimentation and is added to a Krebs-Ringer phosphate buffer solution of the appropriate steroid, plus 15 per cent autologous serum. The steroid solution is prepared by autoclaving the steroid for 60–90 minutes (to inactivate any possible endotoxin), buffer is added and the mixture is shaken overnight. It is then filtered through an ultrafine filter to remove any possible particles. Analyses of all steroids used, by gas-liquid chromatography and melting points, showed that autoclaving did not alter the chemical composition of any steroid. Solutions of heat-labile steroids, such as oestrogen and progesterone, were prepared by dissolving weighed amounts of steroid in sterile, pyrogen-free absolute alcohol and filtering. When used in an experiment, aliquots of steroid in alcohol were added to flasks, the alcohol was blown off with compressed air passed through a sterile pipette, and buffer and serum were added to dry flasks. In these experiments alcohol was also added to control flasks and evaporated. Steroid preparations were always pyrogen-tested by incubating buffer solutions with serum in the same amounts as in experiments with leucocytes, and injecting the solutions into rabbits. The flasks of leucocytes and steroid are incubated for 18 hours at 37°C; the supernatant is then recovered by centrifugation and tested for the presence of endogenous pyrogen by injection into rabbits. Since rabbits do not respond to steroid pyrogens the presence of steroid does not interfere with the assay of endogenous pyrogen.

When leucocytes are incubated with heat-killed staphylococci as a phagocytic stimulus, even one hour of exposure to the stimulating agent is sufficient to ensure maximal release of endogenous pyrogen over the next 18 hours. By contrast, exposure of leucocytes to steroid pyrogen for intervals of less than six hours was insufficient to cause subsequent release of endogenous pyrogen (Table II). Exposure for six or eight hours caused the same stimulation, as measured by the amount of pyrogen released, as did exposure for the full 18 hours.

Studies of the importance of the concentration of steroid in this system indicated that within a very limited range there is a dose response relationship. However, less than 10 μg/ml is usually ineffective in causing any pyrogen release, and more than 20–30 μg/ml appears to be maximally stimulatory.

TABLE II

EFFECT OF LENGTH OF INCUBATION WITH AETIOCHOLANOLONE OR HEAT-KILLED STAPHYLOCOCCI ON 18-HOUR RELEASE OF ENDOGENOUS PYROGEN FROM HUMAN BLOOD LEUCOCYTES

Hours:	0	1	2	4	6	8	20
Stimulus:				*Fever Index*			
Aetiocholanolone	0·06(7)	—	0·14(8)	0·13(6)	0·43(6)	0·45(2)	0·42(8)
Heat-killed staphylococci	0·06(15)	1·05(12)	0·96(15)	—	—	—	0·87(11)

Number of rabbits injected given in parentheses.

Substitution of rabbit leucocytes for human leucocytes in this system prevented release of endogenous pyrogen, even when human serum was used. Thus in the *in vitro* model, as well as in *in vivo* studies, steroid pyrogens appeared to act only on human cells.

When puromycin or actinomycin is added to preparations of leucocytes and pyrogenic steroids, release of endogenous pyrogen does not occur. Thus in this model, as well as in those in which other stimuli to pyrogen release have been investigated (Nordlund, Root and Wolff, 1970), pathways leading to eventual pyrogen production and release apparently involve synthesis of new RNA and protein.

Compared to other activators, such as endotoxin or phagocytosis of heat-killed bacteria, aetiocholanolone caused release of only about half as much endogenous pyrogen from the same number of whole blood leucocytes. On the other hand, when preparations of mononuclear cells from blood were studied, this steroid appeared to cause as much release of pyrogen from these cells as did phagocytosis of bacteria. Very small numbers of blood monocytes ($4–6 \times 10^6$ cells) produced enough pyrogen to cause a rise in temperature of 1°C in a rabbit, whereas ten times this number of whole blood leucocytes usually produced somewhat less endogenous pyrogen. So it is possible that steroid pyrogens act only on the small numbers of monocytes present in the preparation of whole blood leucocytes (Fig. 1). Unfortunately this possibility has not been established with certainty, but no firm evidence has been obtained to suggest that steroids act on polymorphonuclear leucocytes, whereas they clearly do act on monocytes.

The specificity of the chemical configuration required for pyrogenicity *in vivo* has been mentioned previously. It seemed worth determining whether the same specificity exists when human leucocytes are incubated

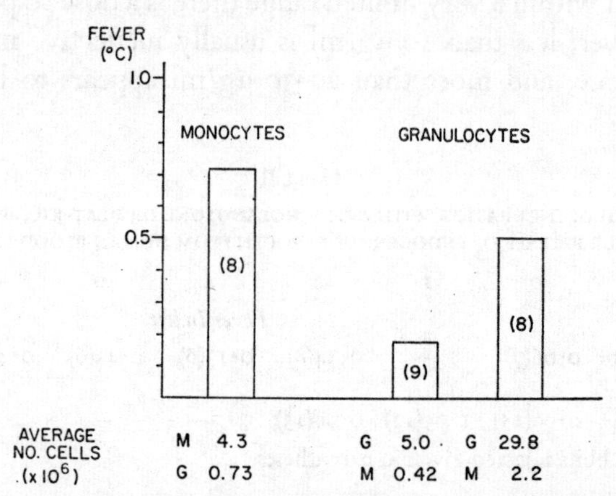

FIG. 1. Average maximum fevers in rabbits following injection of supernatants from 18-hour incubations of white blood cell preparations from normal subjects or patients with agranulocytosis incubated with aetiocholanolone. The average numbers of monocytes (M) and granulocytes (G) contributing to the supernatants are given below each bar. In this and subsequent figures, the number of recipient rabbits is given in parentheses within each bar.

with some of these same steroids. The correlation between the ability to produce fever *in vivo* and to produce endogenous pyrogen *in vitro* is very high, particularly among the C_{19} steroids studied (Table III). Two bile acids also included in Table III, one pyrogenic and one non-pyrogenic *in vivo*, had similar activities *in vitro*. Among the group of C_{21} steroids studied, however, some discrepancies were noted. One possible explanation for these differences would be differences in solubility in buffer solution, since the action of aetiocholanolone is critically dependent on the presence of an adequate concentration. However, both 5β-pregnane-3α,20α-diol and 5β-pregnane-3α-ol-11,20-dione were present in solution in high concentration and were not pyrogenic. Another possibility would be that *in vivo* certain metabolic changes might occur which would not be possible in the *in vitro* incubation system. For instance, oxidation of the 20α-hydroxyl of the

3α,20α-diol would occur readily *in vivo*, and it is possible that the *in vivo* pyrogenicity of this compound may really be due to 3α-ol-20-one, which is pyrogenic both *in vivo* and *in vitro*.

In spite of the potential advantages of working with steroid pyrogens mentioned previously, the sequence of steps involved in release (or synthesis) of endogenous pyrogen has not as yet been clarified by these studies. It appears that steroids, like bacterial or particulate pyrogens (Nordlund, Root

TABLE III

COMPARISON OF *in vivo* AND *in vitro* PYROGENIC EFFECTS OF STEROIDS AND BILE ACIDS

Substance tested	*Concentration in vitro, μg/ml*	*Pyrogenic activity*	
		in vitro	*in vivo*
21-carbon steroids			
5β-pregnane,3α-ol-20-one	8	++	++
5β-pregnane-3α,20α-diol	80	o	++
5β-pregnane-3α-20β-diol	13	+	?
5β-pregnane-3α-ol-11,20-dione	60	o	++
5β-pregnane-3,20-dione	70	++	+
4-pregnene-3,20-dione (progesterone)	26	o	o
5α-pregnane-3,20-dione	5	o	o
5β-pregnane-3α,17α-diol,20-one	60	o	o
5β-pregnane-3β-ol-20-one	29	+	?
19-carbon steroids:			
5β-androstane-3α-ol-17-one	15, 30	++	++
5β-androstane-3α,11β-diol-17-one	12	++	++
5β-androstane-3α-ol-11,17-dione	15, 100	++	++
5β-androstane-3β-ol-17-one	25	o	o
5α-androstane-3α-ol-17-one	30, 91	o	o
5-androstene-3β-ol-17-one (dehydroepiandrosterone)	30	o	o
4-androstene-17β-ol-3-one (testosterone)	5, 25	o	o
Bile acids			
Lithocholic acid	sat. sol'n	++	++
Deoxycholic acid	30	o	o

and Wolff, 1970), require an initial sequence of steps involving synthesis of RNA and protein, but these steps are completed long before the actual release of the pyrogen occurs. They may represent the formation of some activating substance (enzyme?) which acts on a substrate already present in the cell to cause release of the active pyrogen, or an endogenous pyrogen precursor may be newly synthesized. The major difference between the steroids and the other stimuli for pyrogen production is the slow rate of induction of pyrogen in this system. Whether this time is required for early or late steps in cell activation, production, or release of pyrogen is not clear. Presumably the mode of action of the steroid on the cell requires some alteration in cell metabolism, but unlike changes which occur with other activators, no evidence of altered oxygen uptake was found when aetiocholanolone solution was added to whole blood, nor were there detectable morphological changes in the cells. Of course, if only monocytes respond to this steroid, a small effect on respiration would not have been observed. Serum is required for aetiocholanolone to activate blood cells, as is also true for most other cell activators. The inability of these agents to act on species other than man is most puzzling but, as will be mentioned later, this species specificity appears to be true for some effects of oestrogen and cortisol as well.

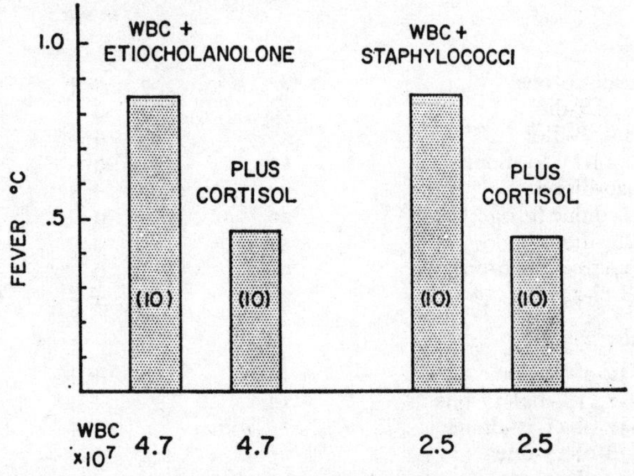

FIG. 2. Average maximum fevers in rabbits following injection of supernatants from 18-hour incubations of human blood leucocytes (WBC) incubated with or without cortisol (12–120 μg/ml) and stimulated by aetiocholanolone or by phagocytosis of heat-killed staphylococci.

Antipyretic effects of cortisone have been known for a number of years. When cortisone was given along with the injection of steroid pyrogen (Palmer and Kappas, 1963) or when subjects were treated with corticosteroid for several days before being given aetiocholanolone (Kappas *et al.*, 1957), the pyrogenic effect of this steroid was blocked. Studies of the antipyretic effect of corticosteroids in rabbits have indicated that their primary effect is probably to influence the response of the hypothalamus (Atkins *et al.*, 1955). Nevertheless, it has been possible to demonstrate *in vitro* that as little as 12 μg cortisol/ml suppresses release of endogenous pyrogen from human blood leucocytes stimulated both by aetiocholanolone and by phagocytosis (Fig. 2). Similar suppression of release of pyrogen has been demonstrated with oestradiol, when phagocytosis was used as the stimulus;

only very little suppression of pyrogen release was observed with aetiocholanolone as stimulus (Fig. 3).

Both cortisol and oestradiol exerted a maximal effect when they were added to leucocytes an hour before the phagocytic stimulus. When either was added to the incubation system one or two hours after the activator, little or no effect was observed. This suggests that both act by altering the early stages of cell activation and production of pyrogen. These steroids do not act by interfering with phagocytosis itself. In addition, if the cells are washed after incubation for 60 or 90 minutes with oestradiol, and then

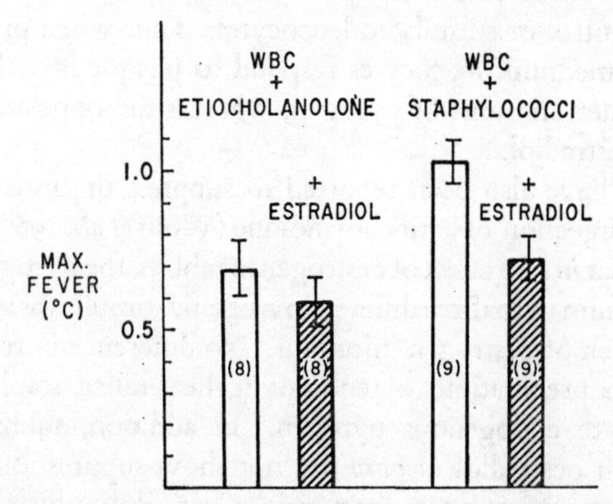

FIG. 3. Average maximum fevers in rabbits following injection of supernatants from 18-hour incubations of human blood leucocytes (WBC) with or without oestradiol (25–30 μg/ml) and stimulated by aetiocholanolone or phagocytosis of heat-killed staphylococci.

stimulated, they behave as do cells which have not been incubated with the hormone. This suggests that the presence of the hormone at the time of stimulation is critical for its effect. The action of oestradiol on pyrogen release varies with its concentration, and a minimal concentration of 10–20 μg/ml appears to be required. These concentrations are similar to those which, in the experiments with aetiocholanolone, are required to elicit pyrogen release. It is perhaps interesting that in other similar experiments using oestradiol, Bodel and co-workers (1970) have demonstrated a variety of effects on phagocytizing leucocytes, including suppression of the normal oxidative burst in metabolic activity, and suppression of the normal amount of granule lysis in polymorphonuclear leucocytes. Thus, in this *in vitro* model, this hormone seems to have a variety of effects on this cell type.

Since progesterone has been noted to cause small sustained increases in body temperature in volunteers, and is presumably responsible for the post-ovulatory rise in basal body temperature in women, it seemed of interest to add progesterone to this system. As mentioned previously, progesterone alone did not cause cells to release pyrogen *in vitro*. Nevertheless, when progesterone-treated cells were stimulated by phagocytosis, augmented pyrogen release could be demonstrated. When both progesterone and oestradiol were used together an intermediate effect was observed, indicating that the two steroids exerted an antagonistic effect in this system. This augmenting action of progesterone on pyrogen release was apparently not due to conversion of progesterone *in vitro* to another steroid, such as pregnanediol, since experiments using tritiated progesterone have demonstrated no conversion of this compound to any other steroid either in the presence of control or stimulated leucocytes. Thus when progesterone is added to the medium, leucocytes respond to phagocytic stimulation by rele asing greater amounts of pyrogen, whereas the opposite effect is observed with oestradiol.

Oestrogens havc also bcen reported to suppress the usual response of subjects to an injection of aetiocholanolone (Wolff *et al.*, 1967). In order to look for a similar *in vivo* effect of oestrogen in rabbits, the pyrogenic response of a group of animals to three different pyrogenic stimuli was studied before and after a week of oestrogen injections. No difference in response could be detected to preparations of endotoxin, heat-killed staphylococci, or rabbit leucocyte endogenous pyrogen. In addition, rabbit leucocytes incubated with oestradiol *in vitro* did not show suppression of pyrogen release after stimulation by phagocytosis. Cortisol, too, was ineffective in this *in vitro* model, even in 100 times the concentration that is effective for human leucocytes. Thus, rabbit leucocytes apparently differ from human leucocytes both in their inability to respond to the pyrogenic steroids by releasing pyrogen, and in the lack of ability of oestrogen and corticosteroids to suppress their normal pyrogen release.

It is natural for a physician to ask whether these effects of oestrogen can be of benefit to patients, in addition to being of interest in experimental studies. We have tried to use oestrogen in five patients who had periodic fevers of long-standing and of unclear origin. In three of these oestrogen administration clearly altered the symptomatology of the disease. One woman, a 26-year-old housewife, had suffered repeated incapacitating attacks of fever associated with peritoneal or pleural inflammation, the so-called familial Mediterranean fever syndrome. She had only a transient improvement when given 50 mg cortisone daily. However, when she was given increasing doses of mixed conjugated oestrogens (mainly oestrone sulphate), she became relatively free of attacks at a dose of 10 mg per day. When the dose was increased by error, she had two "escape" episodes during the subsequent decreasing hormone administration. However, she has since been maintained on 10 or 7·5 mg of hormone daily, and has had only very infrequent attacks on this regimen.

Two patients have been men, with cyclic fevers of predictable onset and duration unassociated with other major symptoms. In these patients oestrogen treatment was given just before the onset of the expected attack, and was continued for one to three days only. With this treatment the attacks were consistently suppressed in both patients, the extent of suppression varying from time to time. No side effects of this treatment were observed. After two years of quite regular monthly attacks, successfully controlled by oestrogen treatment, one patient abruptly stopped having fevers, and has now been asymptomatic for ten months. The second patient has shown alteration in his cycle while on oestrogen treatment, but continues to have regular attacks.

Measurements of plasma aetiocholanolone in all three patients were made at the time of onset of the fever as well as at the height of the fever. In none of these samples were abnormal concentrations of this steroid detected. The possibility that some other steroid pyrogen, or some imbalance of steroid production, is associated with the fevers has not been excluded.

DISCUSSION

The pattern of response of human leucocytes to steroid pyrogens is comparable, in almost every respect, to the response of humans to such pyrogens administered *in vivo*. To be effective, the pyrogens must have certain chemical characteristics; they must be present for a prolonged period as compared with bacterial pyrogens; and they affect human leucocytes but not those of the rabbit. Leucocytes of other species have not yet been studied because the system as used *in vitro* requires that the endogenous pyrogen released by the stimulated leucocytes be effective in an appropriate test animal. Few animals have been shown to have leucocyte pyrogens which produce fever in rabbits. As far as has been determined, the mechanism by which steroid pyrogens affect leucocytes is otherwise entirely comparable to the way in which bacterial or particulate substances operate.

The decision to investigate the effect of oestrogens on pyrogen release was stimulated by the clinical observation that the febrile response to steroid pyrogens is much greater in men than in women (Wolff *et al.*, 1967). Moreover, the syndrome of familial Mediterranean fever is less common in women than in men (Siegal, 1964) although, as indicated by our patient, this syndrome can sometimes be seen in women. The fact that oestrogens are able to antagonize the pyrogenic effects of heat-killed staphylococci to a large extent, but steroid pyrogens only slightly, and that they are effective with human but not rabbit leucocytes is quite peculiar and, as yet, unexplained. Similar oestrogen effects have been reported in other systems. For example, Spangler and co-workers (1969) have shown that oestrogens enhance the anti-inflammatory effect of corticosteroids in treating human skin disease. Moreover, these authors also found that oestrogens alone inhibit granuloma formation induced by carrageenin in guinea pigs, and this effect is synergistic with that of cortisol. The failure of rabbit pyrogen-releasing cells to respond to oestrogens may, therefore, represent a peculiarity of the rabbit rather than a general situation in non-human mammals.

SUMMARY

We have described a system in which the pyrogenicity of steroids can be investigated *in vitro*. Human leucocytes, incubated with certain steroids which cause fever in man, release endogenous pyrogen *in vitro*. The characteristics of this reaction resemble in several ways the pyrogenic response *in vivo* to these same steroids. Both oestrogen and cortisol act on human leucocytes *in vitro* to suppress release of endogenous pyrogen in response to the stimulus of phagocytosis. In addition, in three patients with periodic fevers, oestrogens suppressed or prevented fever and other symptoms. These studies suggest that endogenous steroids in man may play a role in naturally occurring fevers.

REFERENCES

Atkins, E., Allison, F. Jr, Smith, M. R., and Wood, W. B. Jr (1955). *J. exp. Med.*, **101**, 353–366.

Bodel, P., and Atkins, E. (1966). *Proc. Soc. exp. Biol. Med.*, **121**, 943–946.

Bodel, P., Dillard, G. M., Kaplan, S., and Malawista, S. (1970). *J. clin. Invest.*, **49**, 10*a*.

Kappas, A., Hellman, L., Fukushima, D. K., and Gallagher, T. F. (1957). *J. clin. Endocr.*, **17**, 451–453.

Kappas, A., and Palmer, R. H. (1963). *Pharmac. Rev.*, **15**, 123–167.

Nordlund, J. J., Root, R. K., and Wolff, S. M. (1970). *J. exp. Med.*, **131**, 727.

Palmer, R. H., and Kappas, A. (1963). *Med. Clins N. Am.*, **47**, 101–112.

Siegal, S. (1964). *Am. J. Med.*, **36**, 893–918.

Spangler, A. S., Antoniades, H. N., Sotman, S. L., and Inderbitzin, T. M. (1969). *J. clin. Endocr.*, **29**, 650–655.

Wolff, S. M., Kimball, H. R., Perry, S., Root, R., and Kappas, A. (1967). *Ann. intern. Med.*, **67**, 1268–1295.

DISCUSSION

Pickering: You say that the pyrogenic steroids only act in man, so you don't know if they act directly on the hypothalamic centres.

Bondy: I feel that the mechanism by which the pyrogenic steroids act is entirely analogous to the mechanism by which the bacterial pyrogens act except in the length of the latent period and the fact that they seem to affect only human cells. If you put steroids in contact with human cells a substance is released which causes fever in a rabbit; if you put the steroid directly in the rabbit it doesn't cause fever. So I don't think they do act directly on the hypothalamus.

Cranston: Are you entertaining the hypothesis that in women the temperature change at ovulation might be due to a change in the balance of oestrogenic and progesterone type compounds?

Bondy: Yes, but it is not a unique hypothesis. Progesterone in the body is metabolized to pregnanediol which is one of the pyrogenic steroids. In the system we were studying progesterone is not metabolized detectably to pregnanediol even when radioactive tracers are used to follow it.

Cranston: If that hypothesis is correct, then animals which do not respond to these pyrogenic steroids should not have a temperature change when they ovulate, assuming that they do produce progesterone.

Bondy: It would be true if that was the only mechanism of the fever. Rats, for example, have a small rise in temperature every three days in their normal cycle. I suspect that there are several factors involved here. Both progesterone and pregnanediol are relatively poor pyrogens and I think that most of the temperature change in the cycle has to do with changes within the hypothalamus.

Bodel: Although we study an *in vitro* system it doesn't mean that we think the peripheral blood cell is the locus at which these steroids act. It is known for instance that cortisone is antipyretic in rabbits, and appears to act on the hypothalamus (Atkins *et al.*, 1955). Steroids may act at many different places. We find it interesting that we can mimic certain *in vivo* effects using an artificial system.

Cranston: Does oestradiol interfere with phagocytosis?

Bodel: Not that we can detect.

Pickering: Dr Bodel, you said that white cells plus a high concentration of steroids release pyrogen. Does that mean that you have to have a much higher concentration of steroid present than you would expect in the body fluids?

Bodel: Our solution contains about a hundred times the normal plasma concentration of unconjugated steroid. However, the plasma concentration may have little relation to the concentration of a steroid at its active site. We have not investigated binding in our *in vitro* system. Also, I doubt that the blood cells are the cells usually responsible for steroid pyrogenic effects in the body. I think it is more likely that they act, for instance, on Kupffer cells; I just can't study them as conveniently.

Bondy: The most consistent way of producing fever with steroids in humans is to give them intramuscularly. This produces a very extensive inflammatory reaction locally, so that within about an hour, at the point where the injection is made, large amounts of steroid and large numbers of white cells are in close contact. The circulating plasma concentrations may have nothing to do with this artificial situation.

Cranston: You have shown that white cells require a long exposure to aetiocholanolone to produce leucocyte pyrogen and that would fit very well with the duration of infusion that you required.

Bondy: Yes, we are delighted with it.

Cranston: If you give oestrogens to patients with fever due to, say, broncho-pneumonia, does it make any difference?

Bondy: We have not done this very extensively because we do not like giving oestrogens to otherwise normal people.

Bodel: Wolff and co-workers (1967) found that pretreating a small group of women with oestrogen had no effect on bacterial pyrogen (endotoxin) fever, whereas it partially suppressed aetiocholanolone fever. When we gave oestrogen to a patient with periodic fever after his attack had begun it had very little effect, whereas when we started treatment 6–8 hours before the temperature began to rise, his fever was usually suppressed. Perhaps oestrogen has little antipyretic effect once a cell is producing pyrogen, but is able to alter early steps leading to stimulation of that cell.

Cranston: But if that were so and you were able to turn off the production of endogenous pyrogen for a period of 24–48 hours you would expect to be able to show an effect.

Bondy: You are assuming that the fever is due entirely to the endogenous pyrogen.

Bodel: This raises the question of how long a cell produces pyrogen once it is turned on, and how it is turned off. Neither oestrogen nor cortisone seems to act *in vitro* once the cell is turned on. In our usual *in vitro* system, stimulated blood cells only release pyrogen actively for 4–6 hours. Although we do not have good studies on other cell types, such as monocytes, they may follow a different pattern. It is possible that in patients who have a fever lasting two or three days, the stimulus is present for only a brief time, but the cells then release pyrogen for a long time. This could explain the lack of suppression by oestrogen once the fever has begun.

Myers: I was surprised to learn that the monkey does not respond to pyrogenic steroids, because although it is less sensitive to pyrogens than many other animals, it is nevertheless a primate. Could you give us more information about the dose and route of administration of the pyrogenic steroids?

Bondy: Palmer, Ratkovits and Kappas (1961) injected pyrogenic steroids in large doses into *Macaca rhesus* by the intramuscular, intravenous and intrathecal routes without producing fever. It may be that if they had tried chimpanzees they would have had a different kind of response. The dose is probably not critical because in humans whether you give 5 mg or 50 mg in an intramuscular injection, you still get about the same kind of response. Even with 5 mg you are probably giving a supramaximal stimulus.

Work: What is the source of the bacterial pyrogen you use to stimulate your leucocytes?

Bondy: All our studies were with heat-killed staphylococci.

Work: It was mentioned that bacterial pyrogen or staphylococci will give rise to leucocyte pyrogen. If people are just giving vaccines, are they eliminating phagocytosis? Have you any information on the effects of pure lipopolysaccharide on the leucocyte?

Bodel: In most of his studies Dr Atkins used a so-called purified E-pyrogen, made from *Proteus vulgaris*, which is not particulate.

Cranston: Could 5β steroids have a site of action other than the peripheral one? Is there any information about species variability or the likelihood of these steroids crossing the blood-brain barrier?

Bondy: Plasma versus spinal fluid concentration has been studied in the human and the dog and in these two species the permeability and the movement of the steroids seems to be essentially the same (Christy and Fishman, 1961).

Landy: There are two points which I believe could help to relate the work on endogenous pyrogen to that on mediators of cellular immunity. Firstly, it has been well documented that *in vitro* the production of mediators by sensitive lymphocytes responding to antigen can be blocked totally by cortisol; secondly, once these cells are turned on by antigen and are making the array of mediators of cellular immunity, they continue to make them for as long as these cells can be maintained in culture—which is about a week (Lawrence and Landy, 1970). Once turned on, this process continues until the cells disintegrate. I think these are points of similarity that might be worth considering.

REFERENCES

ATKINS, E., ALLISON, F., SMITH, M. R., and WOOD, W. B. (1955). *J. exp. Med.*, **101**, 353.

CHRISTY, N. P., and FISHMAN, R. A. (1961). *J. clin. Invest.*, **40**, 1997–2006.

LAWRENCE, H. S., and LANDY, M. (eds) (1970). *Mediators of Cellular Immunity*. New York: Academic Press.

PALMER, R. H., RATKOVITS, B., and KAPPAS, A. (1961). *J. appl. Physiol.*, **16**, 345–347.

WOLFF, S. M., KIMBALL, H. R., PERRY, S., ROOT, R., and KAPPAS, A. (1967). *Ann. Intern. Med.*, **67**, 1268–1295.

ON THE MECHANISM OF ACTION OF PYROGENS

W. Feldberg

National Institute for Medical Research, Mill Hill, London

When the monoamines noradrenaline and 5-hydroxytryptamine (5-HT) were shown to produce changes in body temperature by their action on the anterior hypothalamus, it was natural to consider the possibility that pyrogens act through release of monoamines. However, this possibility became unlikely when it was found that the monoamines acted differently in different species. For instance, in cats, 5-HT raised temperature and noradrenaline lowered it, whereas the opposite effects were obtained in rabbits, a rise with noradrenaline and a fall with 5-HT. Yet pyrogens raise temperature in both species. So if pyrogens were to act through the monoamines their action would be different in the two species, a release of 5-HT in cats and a release of noradrenaline in rabbits. The real difficulty, however, lay with those species which, like ox and goat, lack a hyperthermic monoamine in the hypothalamus through which pyrogens could act.

There is another reason why it is unlikely that pyrogens act through release of monoamines. The monoamines are released from nerve fibres, monoaminergic fibres, which innervate the cells of the anterior hypothalamus; pyrogens would thus have to act on nerve fibres and not on nerve cells. Pyrogens, apart from producing fever, apparently have no other obvious central effect, so if their action were mediated by the monoamines it would be an action confined to a small group of nerve fibres, those innervating the pre-optic area or the anterior hypothalamus. Such a specific sensitivity is not known in pharmacology for nerve fibres, but it is not unusual for cells of the central nervous system. General pharmacological considerations therefore make it more likely that pyrogens act on cells, and this would exclude an action through the monoamines.

To me, the best evidence against the idea that pyrogens act, or act mainly, through the release of the monoamines, was provided by Cooper, Cranston and Honour (1967). They showed that in rabbits in which noradrenaline raises temperature, depletion of the noradrenaline store in the hypothalamus by reserpine did not prevent pyrogen fever (see Fig. 1). The injection of reserpine into the cerebral ventricles led to a rise in temperature (Fig. 1*a*) and to the disappearance, in the hypothalamus, of the greenish fluorescence characteristic for catecholamines. The rise can be attributed to an effect of the noradrenaline which is released by the reserpine because it did not occur with a second or third intraventricular injection of reserpine given at an interval of 24 or 48 hr, i.e. at a time when no noradrenaline was available for release (Banerjee *et al.*, 1968). Yet in this condition an intravenous injection of pyrogen still produced fever (Fig. 1*b*).

This was a convincing result. But now Teddy (pp. 124–126) has found that

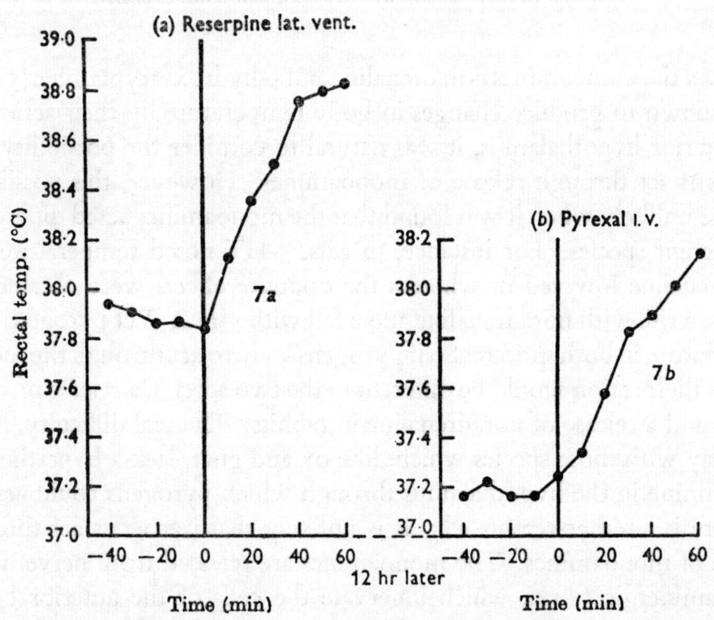

FIG. 1. Rectal temperature in a rabbit showing febrile response to intraventricular reserpine (a) and 12 hours later to intravenous bacterial pyrogen (b). (From Cooper, Cranston and Honour, 1967.)

in rabbits depletion of the 5-HT stores in the hypothalamus by *p*-chlorophenylalanine accentuates pyrogen fever, and depletion of the noradrenaline stores by α-methyl-*p*-tyrosine attenuates it. This certainly suggests an involvement of the monoamines in the fever response. Since the noradrenaline depletion, however, did not abolish but only attenuated the pyrogen fever, it cannot be entirely accounted for by release of noradrenaline. We may therefore still have to look for a monoamine-independent mechanism by which pyrogens act. Myers, Veale and I suggested such a mechanism as the result of an accidental observation (Feldberg, Myers and Veale, 1970).

We wanted to find out whether changes in temperature would result in the release of monoamines into the third ventricle. For this purpose, it became necessary to have a method available for perfusing the cerebral ventricles in the unanaesthetized cat, and we were successful in working out such a method. When the perfusing fluid was artificial cerebrospinal fluid, which contains all the constituents of normal cerebrospinal fluid, body temperature did not change during the perfusion, and it could be continued for an hour or so without the cat showing any signs of discomfort: an affectionate cat showed its affection also during the perfusion, and a restless one had to be lightly restrained. But one day when we used a 0·9 per cent NaCl solution, instead of artificial cerebrospinal fluid, as perfusing fluid, the cat

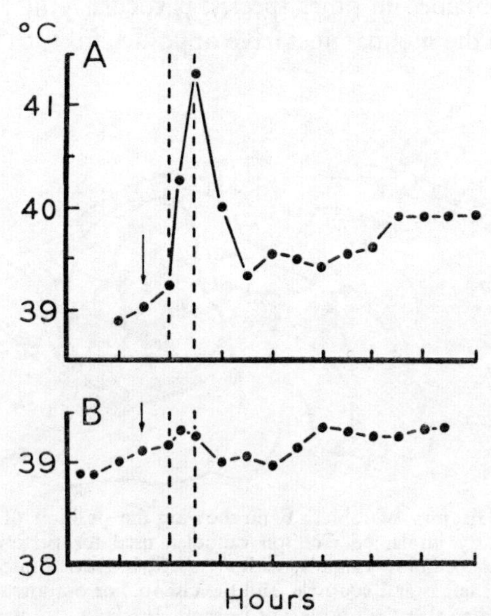

Fig. 2. Record of rectal temperature from a 2·8 kg cat. At the arrows, removal of cat from cage. Between the interrupted lines perfusion from infusion needle, inserted through cannula directly into the left lateral ventricle, to cisterna with a salt solution containing NaCl and KCl (A) or NaCl and $CaCl_2$ (B). (From Feldberg, Myers and Veale, 1970.)

began to shiver vigorously and its temperature rose steeply during the perfusion. Addition of KCl to the NaCl solution did not prevent the rise, but addition of $CaCl_2$ did (Fig. 2).

The perfusion in the unanaesthetized cat was from lateral ventricle to cisterna magna, but the same results were obtained in the anaesthetized cat in which the perfusion was more restricted and was from lateral ventricle to aqueduct (Feldberg, Myers and Veale, unpublished). In these experiments we could further show that the rise in temperature brought about by the lack of calcium was really due to an effect of the sodium ions, because when these were replaced by inert sucrose, and the cerebral ventricles were perfused with an isotonic sucrose solution containing no calcium, temperature did not rise during the perfusion.

We therefore suggested that, in homothermic animals, the constancy of temperature may depend on the balance of sodium and calcium ions in the anterior hypothalamus, that the calcium ions act as a kind of "brake" which prevents the sodium ions from exerting their hyperthermic effect, and also that pyrogens may act by removing this calcium brake. According to this view, pyrogen fever would be a sodium fever.

This view, however, would only be tenable if results obtained in the cat could also be obtained in other species, particularly in a species like the rabbit, in which the monoamines have opposite effects to those in the cat.

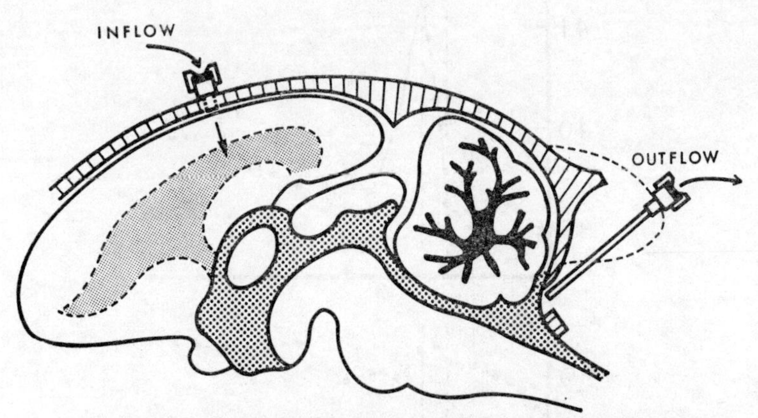

Fig. 3. Diagram of rabbit's brain showing the position of the two permanently implanted Collison cannulae used for perfusion. The ventricular or inflow cannula, without shaft, is screwed into the skull above the left lateral ventricle, and the cisternal or outflow cannula is positioned so that the opening of its shaft lies about 1 mm above the atlanto-occipital membrane. The area between the skull and the interrupted line represents the acrylic cement by which the cisternal cannula is fixed and anchored by two small screws (one being shown) to the back of the skull. (From Feldberg and Saxena, 1970, slightly modified.)

If calcium lack or, rather, the sodium ions, were to raise temperature in rabbits as well, a relatively simple mechanism would be unmasked for raising temperature, independent of the response of a species to the monoamines. This independence is essential for the theory that the constancy of temperature in homothermic animals is determined by the correct balance of sodium and calcium ions in the anterior hypothalamus. And with the same method of perfusion as that used for unanaesthetized cats, Saxena and I obtained in the unanaesthetized rabbit results similar to those previously obtained in cats (Feldberg and Saxena, 1970).

The arrangement for perfusing the cerebral ventricles from lateral ventricle to cisterna magna is shown diagrammatically in Fig. 3. Implantation and fixation of the cannulae was done under anaesthesia, but the perfusions which lasted 20 to 70 minutes were performed without anaesthesia. For each perfusion a needle attached to a slow injector was inserted through the ventricular cannula into the lateral ventricle and the

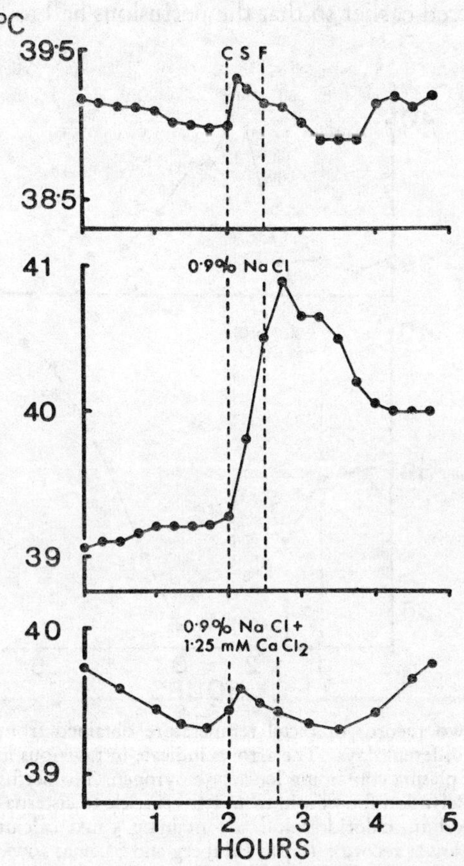

FIG. 4. Three records of rectal temperature obtained from the same rabbit on different days. Perfusion from left lateral ventricle to cisterna for 30 min with artificial cerebrospinal fluid (top record); for 30 min with 0·9 per cent sodium chloride solution (middle record); and for 40 min with 0·9 per cent sodium chloride solution to which 1·25 mM calcium chloride had been added (bottom record). (From Feldberg and Saxena, 1970.)

outflow was collected from a needle which had been passed through the cisternal cannula and pierced the atlanto-occipital membrane.

Figure 4 shows the effect on temperature when the perfusing fluid was artificial cerebrospinal fluid (top record), 0·9 per cent NaCl solution (middle record) and 0·9 per cent NaCl solution to which 1·25 mM $CaCl_2$ had been added (bottom record). This is about the concentration of calcium present in cerebrospinal fluid. The rise in temperature produced during perfusion with 0·9 per cent NaCl solution varied greatly in magnitude in different rabbits. Restlessness, struggling and short-lasting convulsions occurred when these perfusions were continued for over 20 min, but sometimes they occurred earlier so that the perfusions had to be terminated.

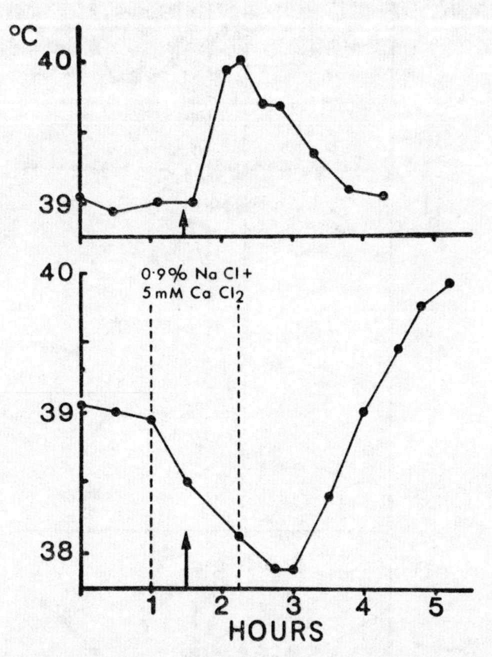

FIG. 5. Two records of rectal temperature obtained from the same rabbit on different days. The arrows indicate intravenous injection of 2·5 ml of plasma containing leucocyte pyrogen. No perfusion (upper record). Perfusion for 75 min from left ventricle to cisterna with a 0·9 per cent sodium chloride solution containing 5 mM calcium chloride (lower record). (From Feldberg and Saxena, 1970.)

We also studied the effect on the rabbit's body temperature of increasing the calcium concentration in the perfusing fluid to 5 mM. This is about four times the concentration in cerebrospinal fluid. In most rabbits such perfusion resulted in a fall in temperature (Fig. 5, lower record), but in some temperature was not affected (Fig. 6, bottom record); for the same rabbit, however, the response was relatively constant on repeated prefusions.

As in cats, replacing the sodium ions by inert sucrose and perfusing the cerebral ventricles with isotonic sucrose solution did not result in a rise of temperature (Fig. 7, lower record).

With these effects established we turned to the next problem. Could the fever produced by the pyrogen be explained by an action on the calcium brake, i.e. by removing the brake? In that case, the pyrogen fever should be prevented either by strengthening the brake, i.e. by perfusing the cerebral ventricles with a solution containing a high calcium content, or by rendering the brake unnecessary, i.e. by replacing the sodium ions with

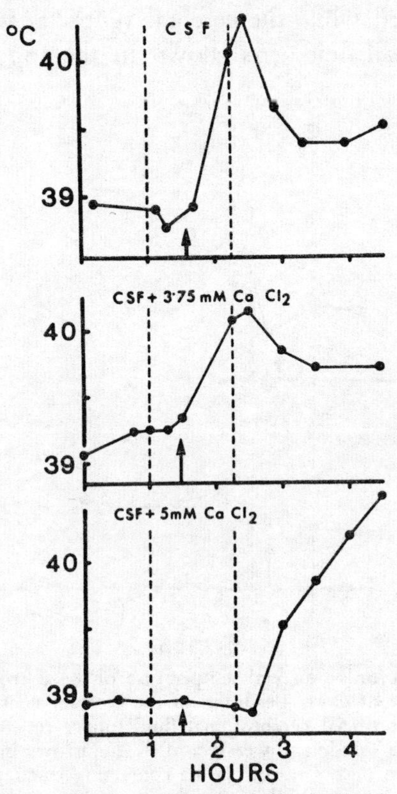

FIG. 6. Three records of rectal temperature obtained from the same rabbit on different days. Perfusion for 75 min from left lateral ventricle to cisterna with artificial cerebrospinal fluid containing different concentrations of calcium chloride. Perfusion with normal artificial cerebrospinal fluid containing 1·25 mM calcium chloride (top record); perfusion with artificial cerebrospinal fluid containing 3·75 mM (middle record) and 5 mM (bottom record) calcium chloride. The arrows indicate intravenous injection of 2·5 ml of plasma containing leucocyte pyrogen. (From Feldberg and Saxena, 1970.)

inert sucrose solution and perfusing the cerebral ventricles with isotonic sucrose solution.

So we examined the effect of perfusing the cerebral ventricles with solutions of different composition on the fever produced by leucocyte pyrogen. The advantage of using leucocyte pyrogen is that approximately the same fever response is obtained when the same dose of pyrogen is injected on different days. The leucocyte pyrogen was prepared by incubating rabbit blood with "E" pyrogen (0·3 μg/100 ml), and 2·5 ml of the plasma containing the pyrogen were injected into an ear vein for 60 to 90 seconds.

First, we found that the fever response was scarcely affected when the pyrogen was injected while the cerebral ventricles were perfused with artificial cerebrospinal fluid. As shown in the top record of Fig. 6,

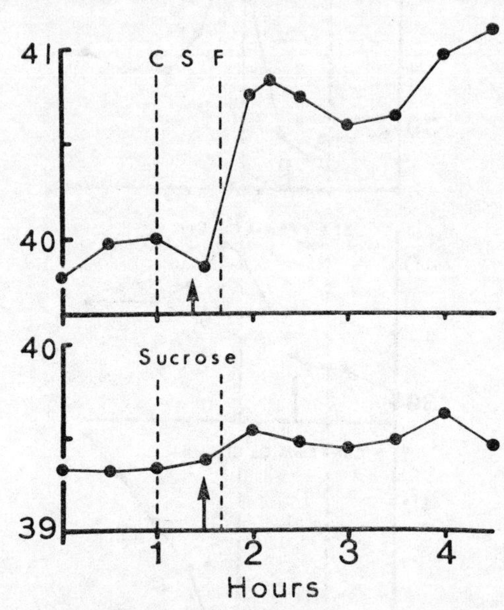

FIG. 7. Two records of rectal temperature obtained from the same rabbit on different days. Perfusion of the cerebral ventricles for 30 minutes with artificial cerebrospinal fluid (upper record) and with isotonic sucrose solution (lower record). The arrows indicate intravenous injection of 2·5 ml of plasma containing leucocyte pyrogen. (From Feldberg and Saxena, 1970.)

temperature began to rise within a few minutes of the pyrogen injection and continued to rise until shortly after the end of the 75 min perfusion. The response was similar to that obtained when the pyrogen was injected into this rabbit without perfusing the cerebral ventricles. Thus the perfusion itself did not affect the pyrogen response.

Next we showed that the response was attenuated or abolished when the pyrogen was injected during perfusion with a solution containing excess calcium. This happened whether such a perfusion had no effect on temperature, as in experiment Fig. 6, or lowered temperature, as in experiment Fig. 5.

The middle record of Fig. 6 shows the attenuated response when the pyrogen was injected during perfusion with artificial cerebrospinal fluid containing about three times the normal calcium, whereas the bottom record shows that the response was abolished when the perfusing fluid contained about four times the normal calcium. The rise seen a short time after the end of this perfusion is not a pyrogen effect, but an as yet unexplained after-effect which occurs sometimes after a perfusion. The lower record in Fig. 5 shows that the fall in temperature produced in this rabbit by perfusion with four times the normal calcium content was not interrupted by the pyrogen injection. The normal response to pyrogen in this rabbit when the cerebral ventricles were not perfused is shown in the upper record.

Thus, strengthening the calcium brake prevents the pyrogen fever, and the same effect was obtained when the calcium brake was rendered unnecessary by perfusing the cerebral ventricles with isotonic sucrose solution (Fig. 7, lower record).

Magnesium ions exerted only a weak calcium-like effect. The hyperthermia produced during perfusion with 0·9 per cent NaCl solution was not prevented by adding 1·25 mM $MgCl_2$ to the solution; even the addition of 5 mM $MgCl_2$ did not abolish the rise though it delayed and greatly reduced it; nor did perfusion with such a solution reduce the pyrogen response. However, an attenuating effect was revealed in the presence of an increased calcium concentration. For instance, in several experiments in which doubling the calcium concentration in the perfusing fluid reduced the pyrogen response, the reduction was greater when the fluid contained 5mM $MgCl_2$ as well.

These are the findings obtained so far. They are in accord with the theory that pyrogen fever is a sodium fever. This does not yet prove that the theory is right. One advantage of the theory however is that it does not use models taken from physics to explain how the set-point works and how it is raised in fever. If the theory is right, it at once poses two pertinent questions not answered by our experiments. First, how does pyrogen remove the calcium brake? Second, is its action confined to the central neuronal mechanisms involved in temperature regulation, or is its action on these neuronal mechanisms part of a generalized action in the central nervous system but with particular sensitivity of the hypothalamic neurons? In this connexion it is necessary to point out that the action of calcium itself, or, rather, of a lack of calcium, is not confined to these neuronal mechanisms. To mention only one example, Merlis (1940) showed in anaesthetized dogs that perfusion of calcium-free salt solutions through the lower spinal subarachnoid space results in augmentation of the flexor reflex, an increase in muscle tone, and spontaneous muscle twitching in the lower half of the body.

SUMMARY

Recent experiments suggest that the constancy of body temperature is determined by the balance of sodium and calcium ions in the hypothalamus, that calcium ions act as a kind of brake and prevent the sodium ions from exerting their hyperthermic effect, and further that pyrogens act by removing the calcium brake. According to this theory pyrogen fever would be a sodium fever. These conclusions are based on results obtained in unanaesthetized cats and rabbits during perfusions of the cerebral ventricles with solutions of different composition.

Temperature did not rise when the perfusing fluid was artificial cerebrospinal fluid, but it rose during the perfusions when the calcium was omitted from this fluid or when the perfusing fluid was simply a 0·9 per cent NaCl solution. The rise was due to the sodium ions because it did not occur when the sodium ions were replaced by inert sucrose. Perfusion with a salt solution containing excess of calcium lowered temperature.

The theory that pyrogens act by removing the calcium brake is based on the findings that strengthening the brake by perfusing the cerebral ventricles with a solution containing excess calcium, or rendering the brake unnecessary by perfusing the ventricles with isotonic sucrose solution, prevented the fever produced in a rabbit by an intravenous injection of pyrogen.

REFERENCES

BANERJEE, U., BURKS, T. F., FELDBERG, W., and GOODRICH, C. A. (1968). *J. Physiol., Lond.*, **197**, 221–231.

COOPER, K. E., CRANSTON, W. I., and HONOUR, A. J. (1967). *J. Physiol., Lond.*, **191**, 325–337.

FELDBERG, W., MYERS, R. D., and VEALE, W. L. (1970). *J. Physiol., Lond.*, **207**, 403–416.

FELDBERG, W., and SAXENA, P. N. (1970). *J. Physiol., Lond.*, **201**, 245–261.

MERLIS, J. M. (1940). *Am. J. Physiol.*, **131**, 67–72.

DISCUSSION

Teddy: One of the problems that has arisen at this symposium is the way in which leucocyte pyrogen, once it is released in the body, brings about a rise in body temperature. Several theories about the mechanisms involved have been put forward; one is that leucocyte pyrogen somehow causes a release of, or alters the concentrations of certain monoamines [5-hydroxytryptamine (5-HT) and noradrenaline] stored in the nerve endings in the hypothalamus. I will describe briefly some experiments which indicate that hypothalamic amines may be involved in the rise in temperature which occurs when leucocyte pyrogen is released.

I investigated the turnover of endogenous amines stored in the nerve endings in the hypothalamus by injecting *p*-chlorophenylalanine, which inhibits the synthesis of 5-HT, and α-methyl-*p*-tyrosine, which inhibits the synthesis of noradrenaline, into rabbits to deplete the monoamine stores in the brain. When there was maximum depletion of the amines stored in the nerve endings, as judged from the results of previous experiments (Giarman

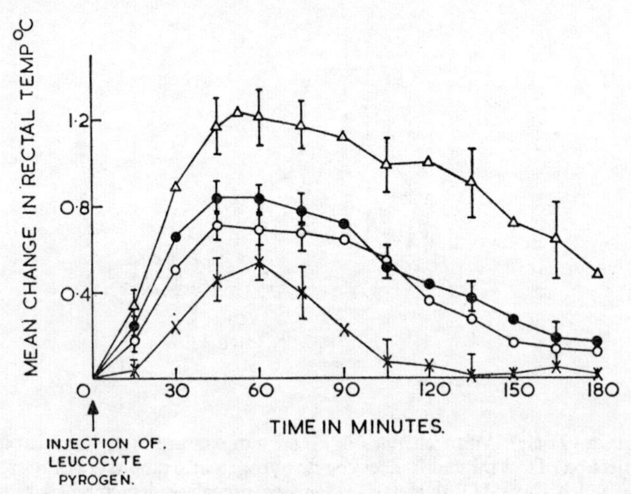

FIG. 1 (*Teddy*). Mean changes in rectal temperature after intravenous injection of 0·1 ml standard leucocyte pyrogen into rabbits (16 or more) previously injected intraperitoneally with the 5-HT depleter or the noradrenaline depleter.
●—● : leucocyte pyrogen after saline (control);
△—△ : leucocyte pyrogen after *p*-chlorophenylalanine;
×—× : leucocyte pyrogen after α-methyl-*p*-tyrosine;
(○—○ : leucocyte pyrogen after α-methyl-*m*-tyrosine).

et al., 1968; Teddy, 1970), I injected crude leucocyte pyrogen into the rabbits and studied the resulting fever curves of the amine-depleted animals and of the control rabbits, which had previously been injected with saline. The maximum rise in temperature and the duration of the fever curve increased in the 5-HT depleted animals, while it decreased in the noradrenaline depleted animals (Fig. 1). These results agree with those obtained in the rabbit by Giarman and co-workers (1968) but they do not really distinguish between the effects of central and peripheral depletion, as noradrenaline and 5-HT are depleted not only in the brain but in the spleen, the blood, the liver and lungs by intraperitoneal administration of the depleting agents. However, when either agent was injected straight into the anterior hypothalamus and then the rabbit was injected intravenously with leucocyte pyrogen at time zero (the time of maximum depletion) the 5-HT-depleted animals again showed a larger rise in temperature after injection of pyrogen and a longer duration of the fever curve than the noradrenaline-depleted animals (Fig. 2). I then injected either the 5-HT depleter or the noradrenaline depleter into the hypothalamus, but the leucocyte pyrogen into the pre-optic area, and again the 5-HT-depleted animal had a larger rise in temperature and a longer-lasting fever curve than the noradrenaline-

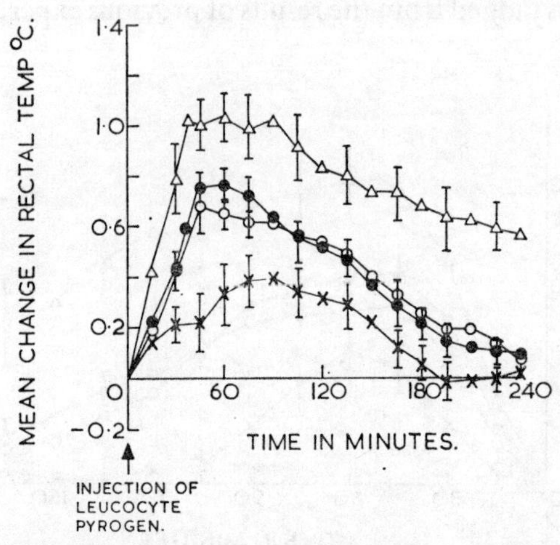

FIG. 2 (*Teddy*). Mean changes in rectal temperature after intravenous injection of 0·1 ml standard leucocyte pyrogen into rabbits (16 or more) in which the 5-HT depleter or the noradrenaline depleter had been previously injected into the anterior hypothalamus.
●—●: leucocyte pyrogen after saline (control);
△—△: leucocyte pyrogen after *p*-chlorophenylalanine;
×—×: leucocyte pyrogen after α-methyl-*p*-tyrosine;
(○—○: leucocyte pyrogen after α-methyl-*m*-tyrosine).

depleted animal (Fig. 3). This suggests that 5-HT and noradrenaline might be involved in central fever pathways. But after injection of either depleting agent into the posterior hypothalamus and injection of leucocyte pyrogen either intravenously or into the pre-optic area, the results were the same as for the control animals. Therefore, there does seem to be some difference between the anterior and posterior hypothalamic mechanisms of temperature regulation.

Pickering: In the experiment shown in Fig. 3 (p. 127) did you inject the depleters and the leucocyte pyrogen through the same needle?

Teddy: No, the arrangement we use, which was originally developed by Dr Cooper, is that a template is attached to the head of a rabbit; from the template, cannulae pass down into the pre-optic area and the anterior hypothalamus. Working with this apparatus I placed the depleters just posteriorly to the area into which the leucocyte pyrogen was injected.

Pickering: Why didn't you put them in the same place?

Teddy: I did try putting them in the same place. The trouble is that one is injecting one microlitre at a time and even two microlitres is a very large quantity to put into the same place in tissue such as rabbit brain.

Saxena: Dr Teddy's experimental results may perhaps be explained on the assumption that the "set-point" in the anterior hypothalamus is dependent on a calcium-sodium equilibrium and its activity is influenced by the amines released locally. The anterior hypothalamus tends to restore towards

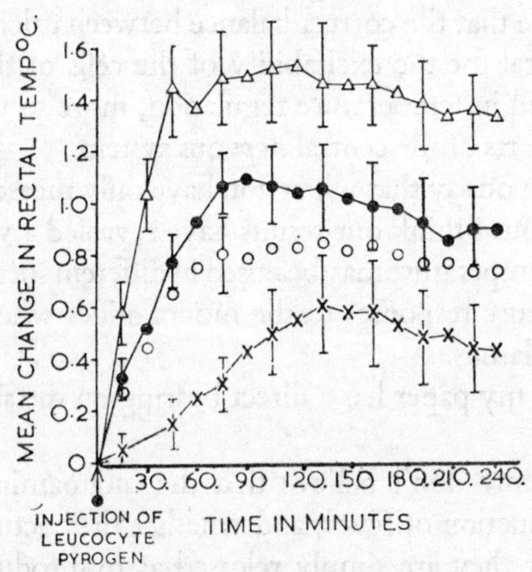

FIG. 3 (*Teddy*). Mean changes in rectal temperature after injection of 1 μl standard leucocyte pyrogen into the pre-optic area of rabbits (16 or more) previously injected with the 5-HT depleter or the noradrenaline depleter into the anterior hypothalamus.
●—●: leucocyte pyrogen after saline (control);
△—△: leucocyte pyrogen after *p*-chlorophenylalanine;
×—×: leucocyte pyrogen after α-methyl-*p*-tyrosine;
(○—○: leucocyte pyrogen after α-methyl-*m*-tyrosine).

normality the temperature effect of leucocyte pyrogens through release of these monoamines. Depletion of the monoamines would then accentuate the effect of the pyrogens.

Cranston: I am not as convinced as Professor Feldberg by our results because our work was not done in a quantitative way (Cooper, Cranston and Honour, 1967). What worries me more, in general, is that if you look at the threshold for the firing of ventricular muscle in an isolated heart, for example, the threshold will be influenced considerably by the calcium and sodium concentrations in the surrounding fluid. If this is so, and if the cells of the central nervous system are functioning in the same way, how do you know that you are not just dealing with a completely non-specific reduction of threshold for firing in all cells? It sounds as if this might be true because the animals had fits, but what is the evidence that this action is in any way related to fever produced by pyrogen? We showed some time ago (Cooper, Cranston and Honour, 1965) that one could produce temperature changes by introducing potassium into the hypothalamus, and we interpreted that as operating in a non-specific way. But why should your observations imply anything more than a general effect on cell membranes of changes in cation concentration?

Feldberg: I am aware that a reduction in calcium has an unspecific effect and increases cell excitability in general. On the other hand I don't find it difficult to assume that the correct balance between calcium and sodium is particularly critical for the excitability of the cells of the anterior hypothalamus involved in temperature regulation, more so than for the excitability in other parts of the central nervous system.

Cranston: Can you say that when you have only measured temperature?

Feldberg: No, but I think our results have revealed a very simple mechanism by which temperature may be raised in different species independently of their temperature responses to the monoamines when they act on the anterior hypothalamus.

Myers: I think my paper has a direct bearing on the site of the sodium-calcium effect.

Teddy: I certainly don't believe that the monoamines are the whole story in the production of fever, and whether they actually help to cause fever or whether they are simply released as by-products during febrile responses, I don't know. If you are going to say that the specific inhibitors of 5-HT and noradrenaline may have other effects, you must admit that reserpine too is a rather inadequate drug for such investigative purposes. Simply measuring brain amine levels is an inadequate way of estimating effects of amines on the central nervous system; it is probably better to do turnover studies with labelled isotopes. I think you are basing some of your earlier assumptions on conclusions drawn from experiments in which reserpine has been used as too fine a biochemical tool, and these conclusions should now be modified.

Feldberg: We felt justified in interpreting the reserpine results of Cooper, Cranston and Honour (1967) differently because of our additional finding that a second or third intraventricular injection of reserpine no longer raised temperature when the noradrenaline stores of the hypothalamus had become depleted. Pyrogen, on the other hand, still worked in this condition, so it seemed to us unlikely that noradrenaline release was responsible for the hyperthermic effect.

Teddy: At the same time you were depleting the 5-HT stores by some 60 per cent.

Feldberg: True, but noradrenaline depletion is greater; it amounts to over 90 per cent.

Teddy: Gross shifts such as these are hard to interpret, and one really needs the results of turnover studies. But the different level is, perhaps, still significant.

Feldberg: The possibility that monoamine release may modify the pyrogen response must certainly be kept in mind, but even in your experiments pyrogen still produced fever after noradrenaline depletion, although the response was attenuated. Was depletion of the noradrenaline stores the only effect of α-methyl-*p*-tyrosine or was there also some change in the sensitivity of the anterior hypothalamus?

Teddy: The effects tended to be dose-dependent. There were no noticeable effects directly attributable to changes in hypothalamic sensitivity until sufficiently large doses were given, which brought about the rapid death of the animal. At the time of death the noradrenaline concentration was usually depleted by about 15 per cent.

Snell: I wish to draw a parallel between these experiments with electrolytes and the findings with activated white cells. Dr Wood showed (Berlin and Wood, 1964) and I have confirmed (unpublished observations) that the electrolytic environment profoundly affects liberation of leucocyte pyrogen by the stimulated cell; potassium in certain circumstances stops it; calcium with one type of stimulus stops it, and with another type of stimulus seems to start it. I don't think that it has been suggested that such environmental electrolyte shifts alone induce release of pyrogen from cells.

REFERENCES

Berlin, R. D., and Wood, W. B. J. (1964). *J. exp. Med.*, **119**, 677.

Cooper, K. E., Cranston, W. I., and Honour, A. J. (1965). *J. Physiol., Lond.*, **181**, 852–864.

Cooper, K. E., Cranston, W. I., and Honour, A. J. (1967). *J. Physiol., Lond.*, **191**, 325–337.

Giarman, N. J., Tanaka, C., Mooney, J., and Atkins, E. (1968). *Adv. Pharmac.*, **6A**, 307–317.

Teddy, P. J. (1970). D. Phil. Thesis, University of Oxford.

HYPOTHALAMIC MECHANISMS OF PYROGEN ACTION IN THE CAT AND MONKEY

R. D. Myers

Laboratory of Neuropsychology, Purdue University, Lafayette, Indiana

Opinions still differ as to how the endotoxin of gram-negative bacteria acts on the central nervous system to produce hyperthermia. Grant, Lewis and Ahrne (1955) reported that the temperature of a rabbit was not altered by injecting endotoxin directly into the anterior hypothalamus. However, it was shown later that *Salmonella typhosa* injected into the cerebral ventricular system caused a reproducible pyrexic response in the dog (Bennett, Petersdorf and Keene, 1957) or cat (Sheth and Borison, 1960).

About ten years ago Villablanca and I confirmed that injecting endotoxin into the cerebral ventricle of a cat caused a high fever, but in a considerably lower dose than when given intravenously. We then began to search for sites in the central nervous system which might be sensitive to bacterial pyrogen (Villablanca and Myers, 1964).

By microinjecting *Salmonella* directly into brain tissue we localized the hyperthermic action of endotoxin to a relatively narrow zone in the anterior hypothalamus. This was accomplished by determining the individual latency of a temperature rise in relation to the distance of the site of the microinjection from the anterior hypothalamus and also by measuring the magnitude of the hyperthermia observed when the pyrogen was given at that site (Villablanca and Myers, 1965). These results were confirmed in the cat by Repin and Kratskin (1967) and Jackson (1967), and in the rabbit by Cooper, Cranston and Honour (1967) who showed that the onset of fever was far more rapid if leucocyte pyrogen was injected into the anterior hypothalamus. The volumes of endotoxin and leucocyte pyrogen used by Cooper and his colleagues were far less than those used by Grant, Lewis and Ahrne (1955) and this could explain the discrepancy in their reports.

PYROGEN AND THE ANTERIOR HYPOTHALAMUS

It is uncertain whether in man the bacterial endotoxins act directly on the central nervous system to produce fever (Cooper, 1965). Because of its phylogenetic closeness to man we wondered about the effects of pyrogen on the monkey. Although some of the infra-human primates are relatively unresponsive to pyrogens (Rašková and Vaněček, 1964), monkeys and apes have been used for a variety of investigations into the problems of pathogens, infectious disease and the pattern of a febrile response. In the *Pongidae* an endotoxin, such as *S. typhosa*, given systemically exerts a powerful haematological effect and causes fever to develop. In eight chimpanzees Tully, Gaines and Tigertt (1965) found that in addition to leucopenia and other clinical signs, the average rise in temperature began within the first hour and rose by as much as 1·5°C after an intravenous injection of 1·3 μg of *S. typhosa*. Generally, the chimpanzees did not shiver and they required a dose of endotoxin 40 times greater than that used to produce a comparable effect in the human.

"Driving" the thermostat of the monkey with endotoxin

The rhesus monkey seems to be even more unresponsive to systemically administered endotoxins than his ape cousin. In the immature *Macaca mulatta* intravenous injections of 10 or 12 mg *S. typhosa*/kg body weight produce little if any temperature response unless the monkey is tightly restrained in a supine position and covered with a light blanket (Sheagren, Wolff and Shulman, 1967). A dose-response relationship could not be established with these extraordinarily high amounts which caused an average rise in temperature of only 0·8°C. Although these results are difficult to explain, restraint increases the monkey's activity, a blanket prevents heat dissipation, and therefore a threshold response to the endotoxin would occur. Moreover, in the young monkey some residual antibody protection might influence the response to a pyrogen.

We have found that the rostral part of the hypothalamus of the monkey is particularly sensitive to the presence of a pyrogen, in much the same way as that of the cat or rabbit, and we have used *Escherichia coli*, *Shigella dysenteriae* and *S. typhosa* to "map" large areas of the diencephalon and mesencephalon of relatively mature unanaesthetized rhesus monkeys (Myers, Rudy and Yaksh, 1971).

The animal is fully acclimated to a primate-restraining chair before the stereotaxic implantation of the cannula guide. The temperature of the venous blood, measured from the sagittal sinus, closely parallels the colonic temperature which is usually monitored simultaneously.

Microinjecting 0·5 to 1·2 μl (see Fig. 1) of a bacterial endotoxin in a concentration of 1×10^8 organisms/ml into the anterior hypothalamic pre-optic region caused a long-lasting fever to develop. Both the concentration of the bacterial organisms in the hypothalamus and the specific

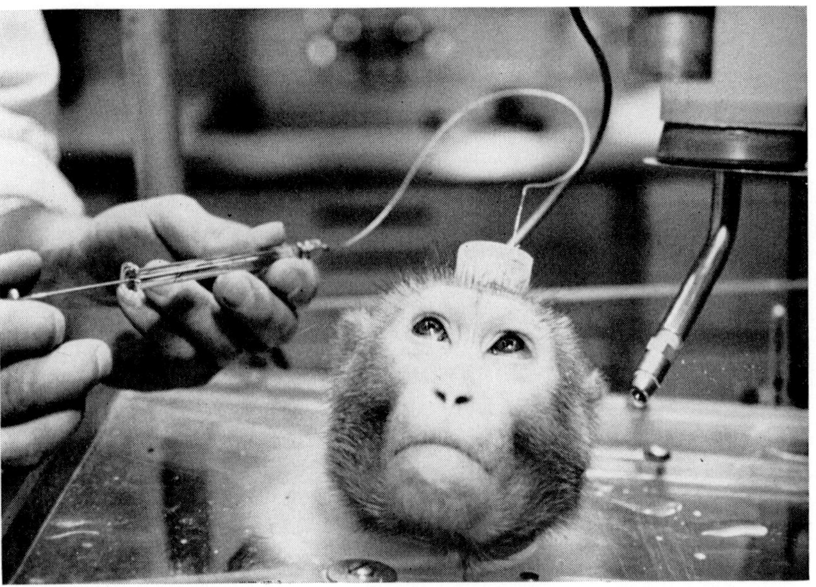

Fig. 1. Microinjection of an endotoxin into the anterior hypothalamus of a rhesus monkey which is fully acclimated to a primate-restraining chair before surgery. Temperature responses are monitored from a thermistor bead implanted against the sagittal sinus.

(*Facing page 132*)

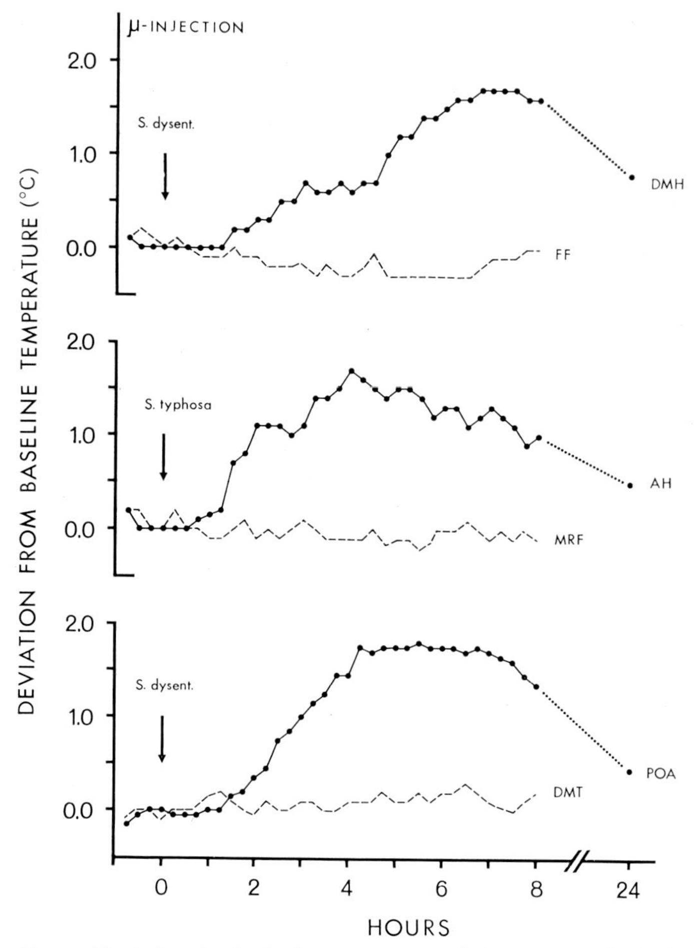

FIG. 2. Deviations in the body temperature of three unanaesthetized rhesus monkeys in which 1·2 μl of a pyrogen were injected at zero hour in a 1:2 dilution at three-day intervals. *Top*: *Sh. dysenteriae* microinjected in the dorso-medial hypothalamus (DMH) and fields of Forel (FF). *Middle*: *S. typhosa* microinjected in the anterior hypothalamus (AH) and mesencephalic reticular formation (MRF). *Bottom*: *Sh. dysenteriae* microinjected into the pre-optic area (POA) and dorso-medial nucleus of the thalamus (DMT). (From Myers, Rudy and Yaksh, 1971.)

anatomical site at which the endotoxin is microinjected are important factors in the development of the fever. Fig. 2 illustrates the pyrexic response to 1·2 μl of *Sh. dysenteriae* or *S. typhosa* injected into different regions of the hypothalamus, mesencephalon or thalamus of three unanaesthetized rhesus monkeys. *Sh. dysenteriae* given in the dorso-medial hypothalamus (Fig. 2, *Top*) caused a long-lasting slow rise in temperature after a latency of over one hour. When *S. typhosa* was microinjected into the anterior hypothalamus (Fig. 2, *Middle*) the temperature rose far more rapidly and the latency was usually less than 30 minutes. Similarly, endotoxin given in the pre-optic area (Fig. 2, *Bottom*) caused a long-lasting rise in temperature with a latency and rate of rise approximately equal to that produced by injecting into the anterior region. In each of these experiments the monkey exhibited all the signs of mobilized heat production including piloerection, vasoconstriction (as evidenced by cold ears and a blanched face) and intermittent shivering, particularly during the steepest portion of the fever curve. At the plateau shivering seemed to occur only in bursts.

Microinjections of endotoxins at other sites in the brain-stem evoke little if any change in the monkey's temperature. The same dose of *Sh. dysenteriae* (1·2 μl of a 1 : 2 dilution of 1×10^8 organisms/ml normal saline) was given in the first monkey at a site close to the fields of Forel and adjacent to the zona incerta (Fig. 2, *Top*). In the second monkey, *S. typhosa* was microinjected in the mesencephalon in a region bordering the red nucleus and the interstitial nucleus of Cajal (Fig. 2, *Middle*); and in the third monkey, the *Sh. dysenteriae* was injected in the dorso-medial nucleus of the thalamus (Fig. 2, *Bottom*). So far, Rudy, Yaksh and I have not been able to find other areas in the basal ganglia or brain-stem of the primate which consistently mediate a febrile response after the local application of an endotoxin. This is surprising as other investigators have found that in rabbits pyrogen can act on certain areas in the mesencephalon, although the latency of pyrexia is apparently greater than that following microinjection of the same endotoxin into the pre-optic area (Rosendorff, Mooney and Long, 1970).

Tachyphylaxis to central pyrogen

One of the most remarkable features of the centrally induced pyrexic response is the absence of tachyphylaxis after repeated microinjections of an endotoxin. Villablanca and I (1965) found this in the cat, and more recently Rudy, Yaksh and I have witnessed 16 successive fevers of equal magnitude after injecting *E. coli* (W3110 strain) into the anterior hypothalamus (Myers, Rudy and Yaksh, 1971). The experiments were carried out over 34 days at two- to five-day intervals. This observation is remarkable when one considers the rapid onset of tachyphylaxis observed with several intravenous injections of an endotoxin (Atkins, 1960).

At intervals of 72 hours 1·2 μl of a 1:2 dilution of 10^8 *E. coli* (W3110 strain) organisms/ml normal saline were microinjected into the anterior hypothalamus of an unanaesthetized rhesus monkey, and it is readily seen from Fig. 3 that on each occasion the intensity of the hyperthermia was very similar within three hours of the injection.

These results indicate that each time a bacterial substance is injected into the pyrogen-sensitive zone of the hypothalamus the microorganisms exert a profound effect on the cells of that region, and also that no local mechanism apparently exists in the rostral hypothalamus to reduce the sensitivity of these cells to the pyrogen; thus tachyphylaxis does not occur.

Salicylate and hypothalamic fever in the monkey

How an anti-pyretic modifies the fever produced by a bacterial pyrogen is still unresolved. At present several views are held. One is that salicylate may interfere with the release of leucocyte pyrogen from cells, since the compound reduces the yield of pyrogen when leucocytes are incubated with endotoxin (Gander, Chaffee and Goodale, 1967). Another view is that a salicylate could attenuate a pyrogen fever by limiting the passage of leucocyte pyrogen into the pyrogen-sensitive areas of the rostral hypothalamus (Cooper, Grundman and Honour, 1968). In rhesus monkeys a salicylate evoked a dose-dependent defervescence after the animal's fever had reached a plateau, which is similar to what is seen in man (Adler *et al.*, 1969). When Rudy, Yaksh and I microinjected 1·2 μl of *E. coli* into a site in the anterior hypothalamus this caused a fever in a dilution as low as 1:2000. Fig. 3 presents the results of three experiments carried out at 72-hour intervals. On each occasion, after the fever had reached a plateau, 300, 600 or 1200 mg of sodium salicylate were given intragastrically via a naso-pharyngeal tube in 25 ml warmed saline. The highest dose (Fig. 3, *Bottom*) abolished shivering, caused transient vasodilation and seemed to make the monkey quiescent. At the same time the temperature declined and remained lower for nearly four hours, which was approximately twice as long as after the administration of 600 mg sodium salicylate (Fig. 3, *Middle*). The lowest of these three doses was apparently at the threshold of the effect of the drug (Fig. 3, *Top*).

These results suggest that the salicylate acts centrally on those cells of the anterior pre-optic region to which the local application of an endotoxin evokes a pyrexic response. Further, these findings support those of Wit and Wang (1968) who showed that the discharge rate of warmth-sensitive neurons, which had been suppressed by endotoxin given intravenously, was reinstated after the injection of acetylsalicylate into the carotid artery. Other cells did not exhibit these changes in response to the salicylate, and

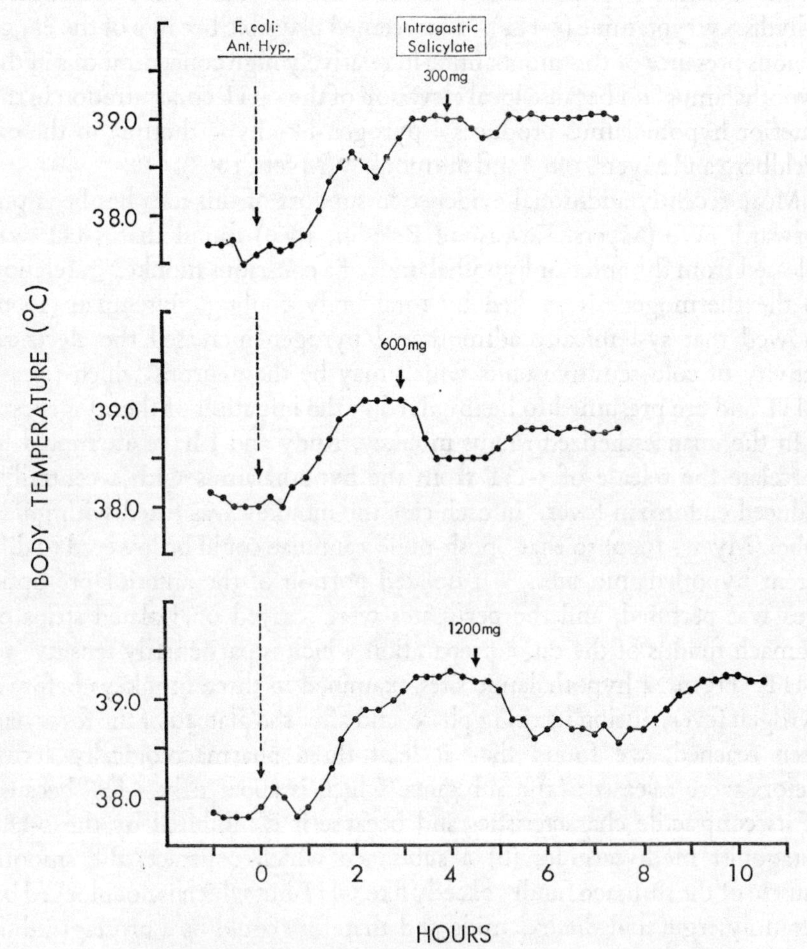

FIG. 3. Fevers produced in one unanaesthetized rhesus monkey by 1·2 µl injections of 1 : 2 dilution of *E. coli* into the anterior hypothalamus (Ant. Hyp.) at three-day intervals. Sodium salicylate (300, 600 or 1200 mg) was given to the 6 kg monkey by the intragastric route at the time indicated by each arrow. (From Myers, Rudy and Yaksh, 1971.)

in our experiments with monkeys there was no circulating pyrogen in the plasma. Hence, it would appear that a salicylate acts in some direct or indirect way on the hypothalamus. This does not rule out the possibility that the drug may also affect the manufacture of leucocyte pyrogen or modify its entrance into the diencephalon through the blood-brain barrier.

Pyrogen fever and 5-HT release from the hypothalamus

In 1965 Villablanca and I suggested that pyrogen might cause the cells of the anterior hypothalamus to release the candidate neurotransmitter 5-hydroxytryptamine (5-HT). This seemed plausible because of the endogenous presence of this monoamine in relatively high concentrations in the hypothalamus and because local elevation of the 5-HT concentration in the anterior hypothalamus produces a pyrogen-like hyperthermia in the cat (Feldberg and Myers, 1965) and the monkey (Myers, 1968).

More recently additional evidence in support of this idea has been put forward. We (Myers, Kawa and Beleslin, 1969) found that 5-HT was released from the anterior hypothalamus of a conscious monkey in relation to the thermogenesis evoked by total body cooling. Eisenman (1969) showed that systemically administered pyrogen increased the electrical activity of cold-sensitive units which may be the neurons which release 5-HT and are presumed to be involved in the initiation of thermogenesis.

In the unanaesthetized rhesus monkey, Rudy and I have attempted to correlate the release of 5-HT from the hypothalamus with a centrally-induced endotoxin fever. In each case the monkey was fitted with guide tubes (Myers, 1970) so that "push-pull" cannulae could be lowered to different hypothalamic sites. An isolated portion of the anterior pre-optic area was perfused, and the perfusates were assayed on isolated strips of stomach fundus of the rat, a preparation which is particularly sensitive to 5-HT. From 24 hypothalamic sites examined in three monkeys before a pyrogen fever, during its rising phase, and after the plateau of the fever had been reached, we found that at least three pharmacologically active factors were released: (a) a substance which is apparently 5-HT because of its contractile characteristics and because it is inhibited by the 5-HT antagonist methysergide; (b) a substance which contracts the smooth muscle of the stomach fundus exactly like 5-HT but which is not blocked by bromolysergic acid diethylamide and therefore could be a prostaglandin; and (c) a third substance which relaxes rather than contracts the stomach fundus strip and whose properties are reminiscent of a catecholamine.

Effluent collected from 17 of the sites that we perfused released one or more of the three factors, either during the rising phase or after the fever had attained a plateau. About 50 per cent of the sites released a 5-HT-like substance, but half of these samples were methysergide-resistant; the other 50 per cent released the unknown relaxing substance. Although the cells at these sites are capable of releasing 5-HT during thermogenesis, their role in the activation of a pyrexic response is not yet clear (Myers and Beleslin, 1971).

PYROGEN AND THE POSTERIOR HYPOTHALAMUS

The above has shown that pyrogens have a profound local action on the cells of the anterior hypothalamus, a region which is generally acknowledged to contain the thermostat (Hardy, 1961), or, as it has been more recently called, the "chemical thermostat" (Myers, 1969). After the early study of Ranson, Clark and Magoun (1939) on the effects of widespread diencephalic ablations on a pyrogen-induced fever, Thompson, Hammel and Hardy (1959) showed that in dogs with partial lesions of the posterior hypothalamic grey matter, a pyrogen produced only a moderate hyperthermia. When the caudal area was entirely destroyed endotoxin given intravenously failed to produce a fever. The posterior hypothalamus in its traditional role as the heat maintenance "centre" (Benzinger, 1964) would thus seem to play a vital role in the genesis of a pyrogen-induced fever. Moreover, the concept that a pyrogen shifts the set-point for body temperature upwards (Fox and MacPherson, 1954; Wit and Wang, 1968; Eisenman, 1969) was early established by Andersen, Hammel and Hardy (1961) when they demonstrated that cooling the anterior hypothalamus of the dog produced a "hyper-fever", while warming it produced a "hypo-fever" or completely prevented the development of pyrexia. It should be emphasized that both of these responses are related directly to the length of time during which the anterior hypothalamus of the dog was either heated or cooled.

Ionic alteration of the set-point

As described by Professor Feldberg (1971), we recently showed that a solution of isotonic sodium chloride caused shivering and a rise in temperature when perfused through the cerebral ventricles of the conscious cat (Feldberg, Myers and Veale, 1970). Addition of calcium in a normal physiological concentration to the NaCl solution blocked the hyperthermic response. We therefore suggested that the calcium concentration in the hypothalamus may be the physiological basis of the set-point. However, sodium itself may be the active ion involved in heat maintenance, and Veale and I tested this by perfusing an isotonic sucrose solution from the lateral ventricle to the aqueduct in two unanaesthetized cats. To our surprise the temperature of the cats did not change (unpublished observations, 1969).

Although the caudal region of the hypothalamus is insensitive to the application of 5-HT, catecholamines and pyrogens, body temperature can be altered markedly simply by shifting the balance of essential cations within the posterior hypothalamus. From our experiments with cats,

Veale and I drew three conclusions about the maintenance of the body temperature of a mammal at or about 37°C throughout life: (1) the mechanism for the temperature set-point lies in the posterior hypothalamus rather than in the anterior pre-optic region; (2) sodium ions are as important as calcium ions in the ultimate determination of a stable set-point; and (3) the ratio of these two essential cations in extracellular fluid, rather than the

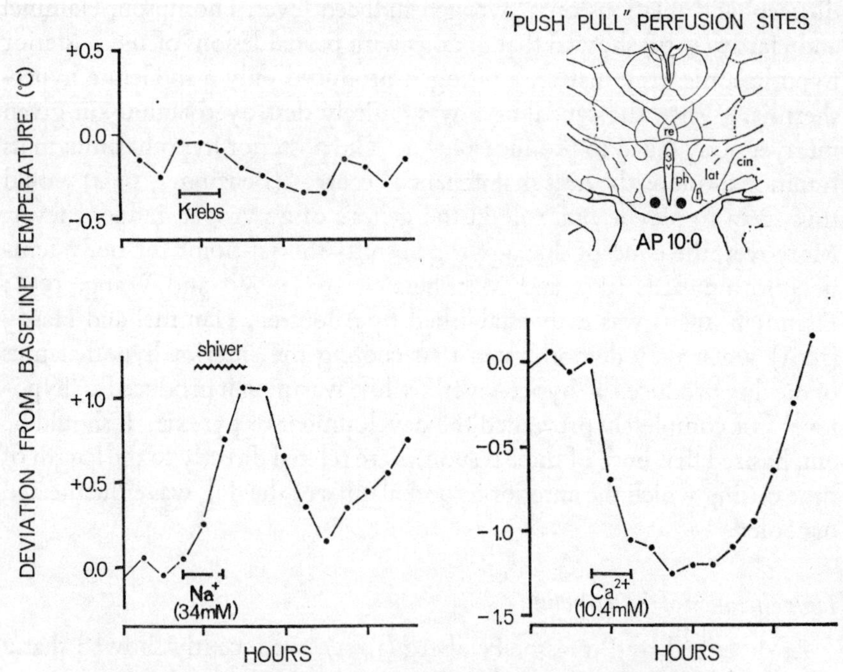

FIG. 4. Changes in the colonic temperature of an unanaesthetized cat in response to the local perfusion of the posterior hypothalamus for 30 min with a Krebs solution alone (*upper left*); Krebs solution plus 34 mM excess sodium (*lower left*); Krebs solution plus 10·4 mM excess calcium (*lower right*). The site of each bilateral perfusion is designated by the dots in the inset. Shivering occurred as indicated. (Modified after Myers and Veale, 1970.)

specific level of either one, is what governs the steady-state discharge pattern of cells to set the temperature at a specific point (Myers and Veale, 1970).

In unanaesthetized cats chronic guide tubes were implanted just above the rostral or caudal portion of the hypothalamus according to procedures described earlier (Myers, 1970). Perfusions of isolated sites within these two regions were made possible by lowering the "push-pull" cannulae through the guide tubes. Fig. 4 illustrates the changes in the body temperature of the conscious cat after sodium or calcium (both ions were in excess of normal physiological concentrations), or normal Krebs solution were perfused at 50 μl/min at the sites shown in the inset of Fig. 4. The slight rise in sodium concentration in these regions produced a rapid rise in temperature which continued even after the 30-minute perfusion interval, but a slight excess in calcium ions evoked hypothermia. From these results, Veale and I believe that the inherent ratio between the sodium and calcium ions probably determines the steady-state activity of nerve cells within the posterior hypothalamus (Myers and Veale, 1971). A slight deviation in the ratio of these ions would affect the depolarization of the neurons,

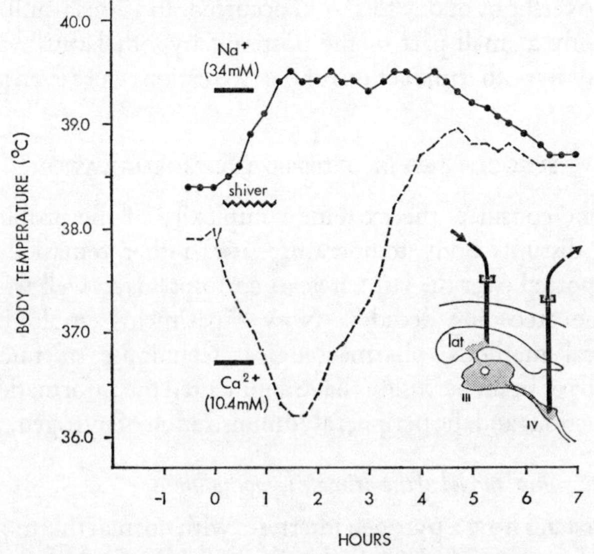

FIG. 5. Changes in the temperature of an unanaesthetized rhesus monkey in response to the ventricular perfusion (0·1 ml/min for 40 min, at 48-hour intervals) of artificial cerebrospinal fluid plus either 34 mM excess sodium (*upper trace*) or 10·4 mM excess calcium (*lower trace*). Inflow was in the lateral ventricle (lat) and effluent was collected from the fourth ventricle (IV) as shown in the inset.

which is in accord with the general suggestion made by Cooper (1965) that a pyrogen may alter depolarization characteristics.

The brain-stem of the primate is just as sensitive as that of the cat, or even more so, to slight changes in concentrations of essential cations. When a Krebs solution or an artificial cerebrospinal fluid was perfused through the ventricular system of an unanaesthetized monkey its temperature did not change. However, perfusion with sodium in slight excess of the normal extracellular fluid concentration caused a sharp increase in temperature; excess calcium in the perfusate produced hypothermia (Myers and Veale, unpublished observations). In these experiments, the sterile solutions were perfused for 40 minutes from the lateral ventricle to the fourth ventricle (see Fig. 5, inset) in which modified Collison cannulae had been implanted. Fig. 5 shows the typical hyperthermia produced by perfusing sodium ions in excess of normal by 34 mM at 0·1 ml/min. During the rising phase the animal shivered intensely. Ordinarily the temperature returned to the baseline more rapidly in other experiments but the extracellular excess in sodium ions, in this case, probably did not dissipate so rapidly. Calcium, 10·4 mM in excess of the normal value, produced a sharp fall in temperature which persisted for a short time even after the perfusion ceased. When the temperature returned to normal, usually an overshoot of 0·5 to 1·0°C occurred. In a "push-pull" perfusion, in which only a small part of the posterior hypothalamus was perfused, the hyper- or hypothermic actions of the two cations were even greater.

PYROGEN AND DISORDERED THERMOREGULATION

When one considers the extreme complexity of the problem of how a pyrogen disrupts body temperature, it is rather remarkable that the research reported over the last ten years corresponds so well with the literature of the preceding decades. Now experiments employing electrophysiological methods, pharmacological techniques of microinjection and methods of brain perfusions have augmented the information obtained with brain lesions and the peripheral administration of pyrogen.

Temperature control model of the primate hypothalamus

To understand how a pyrogen interferes with normal thermoregulation, a theoretical model of the hypothalamic control mechanisms in the primate is presented in Fig. 6. Each component of this model of transmitter and ionic function has been derived largely from physiological and pharmacological experiments performed with the rhesus monkey and from some comparative evidence obtained with the cat. Parts of the model, therefore, should be considered as provisional, although a neurochemical continuity seems to be maintained phylogenetically from the cat to the monkey.

Thermal information is supplied to the anterior pre-optic area of the hypothalamus presumably from two sources: (1) the temperature of the blood (Hayward and Baker, 1968), and (2) the afferent neural impulses from the periphery (Iggo, 1969). Individual 5-HT and noradrenaline-containing neurons follow an ascending course through the mesencephalon (Fig. 6), and terminate in the hypothalamus. The 5-HT sensitive cells and the cold-sensitive neurons apparently are one and the same; the noradrenaline-sensitive cells and the warmth-sensitive units also appear to be identical (Beckman and Eisenman, 1970). From the results of the experiments in which 5-HT was either microinjected or was found to be released in hypothalamic perfusates, it is envisioned that (a) the arterial blood that is lower in temperature than the normal set-point temperature and (b) the afferent impulses arising from peripheral cold receptors serve to activate the 5-HT-containing cells. Then 5-HT is released into the synapse to trigger the efferent pathway for heat production.

This thermogenic pathway with its origin in the anterior region seems to be mediated through a cholinergic system of cells. This concept is based

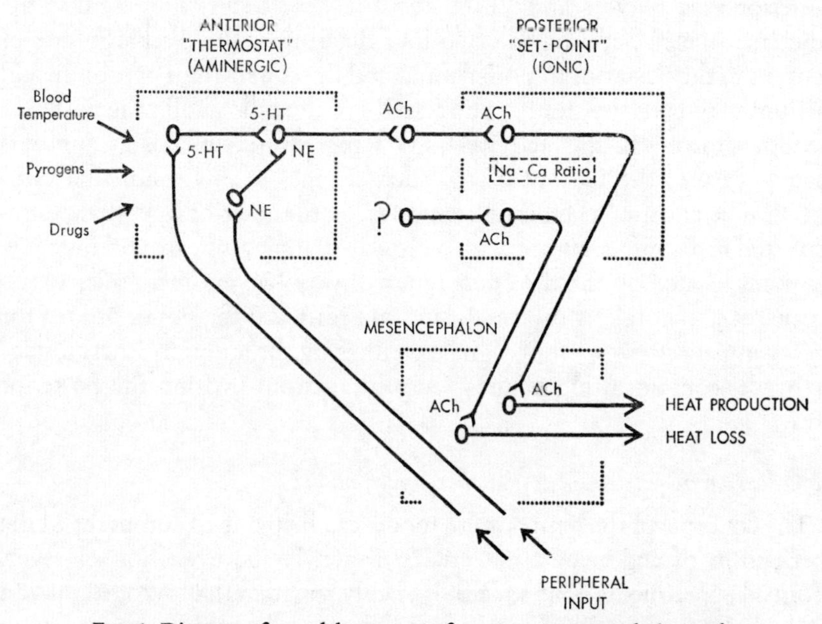

FIG. 6. Diagram of a model to account for temperature regulation under normal conditions as well as during a pyrogen fever. Factors which affect the aminergic "thermostat" in the anterior hypothalamus are given, and the outflow to the posterior hypothalamic "set-point" is mediated by a cholinergic system which passes through the mesencephalon. (See text for a fuller explanation.) 5-HT: 5-hydroxytryptamine; NE: noradrenaline; ACh: acetylcholine.

on two findings: (1) acetylcholine applied to more posterior sites in the hypothalamus produces hyperthermia (Myers and Yaksh, 1969); and (2) acetylcholine is released just caudal to the anterior hypothalamus when a monkey is cold (Myers and Beleslin, 1971). When warm blood or a chemical agent such as an anti-pyretic or anaesthetic touches the catecholamine-containing cells, noradrenaline is released onto post-synaptic sites to block the cholinergic heat production pathway; in this way, the catecholamine serves to activate indirectly a secondary cholinergic system which arises at the junction of the hypothalamus and mesencephalon and which mediates the heat loss system (Myers and Yaksh, 1969).

In addition to the cholinergic synapses for heat production and dissipation, the posterior hypothalamus, traditionally regarded as the heat maintenance centre, apparently contains the cells responsible for maintaining the temperature set-point. Although the caudal region is insensitive to the application of a pyrogen, 5-HT or a catecholamine, it was shown in the previous section how body temperature can be altered markedly simply by shifting the ratio between sodium and calcium ions within the posterior area (Myers and Veale, 1970). These results suggest that the inherent balance between the two ions determines the discharge rate of neurons in the posterior hypothalamus, and the equilibrium of sodium and calcium maintains the steady-state activity of the cells within the posterior hypothalamus. By this activity body temperature remains at approximately 37·0°C throughout life. When changes in environmental temperature occur, the "chemical thermostat" in the anterior hypothalamus is activated in a compensatory way to regulate temperature around the ionic set-point located on the direct efferent pathway (Myers and Veale, 1971). Again, acetylcholine is the candidate transmitter in the posterior area for these temperature responses, signalled either by the anterior "thermostat" or by the constant firing of the "set-point neurons" within the posterior area.

How does a pyrogen disrupt normal thermoregulation?

The concepts of the temperature model can help one to comprehend the mechanism of action of a bacterial pyrogen in the evocation of fever. From all experiments done so far it is readily apparent that pyrogens have a dual action on the cells of the hypothalamus. A bacterial agent exerts an effect on the cells of the anterior hypothalamus which is different from its action on the cells of the posterior region. Thus a febrile response is produced by the combined interference with the normal function of the cells of the "anterior thermostat" and the "posterior set-point". Table I presents a summary of the independent actions of a pyrogen on the anterior and posterior hypothalamic areas.

Firstly, as cited in the previous section, 5-HT is released from the anterior hypothalamus of a monkey exposed to a cold environment; however, it is still uncertain whether it is released from the rostral area of the hypothalamus to trigger the fever brought about by a pyrogen. Nevertheless, the evidence favouring the involvement of 5-HT in the thermogenesis resulting from a pyrogen is reasonably strong. Microinjecting both endotoxin and 5-HT into the anterior pre-optic area of a monkey produces hyperthermia (Myers and Yaksh, 1969; Myers, Rudy and Yaksh, 1971). The similarity of the pyrogen fever and 5-HT hyperthermia may be the key to understanding the local action of a bacterial agent. Humphrey and Jaques (1955) showed that an endotoxin causes 5-HT to be liberated from platelets; the same response could occur within the capillary beds of the anterior hypothalamic microcirculation. The 20- to 40-minute latency of the fever observed after a pyrogen is injected into the anterior hypothalamus could be explained on the basis that the local concentration of 5-HT, originating from the platelets, must exceed a threshold concentration in order to act on 5-HT receptor sites and evoke the pyrexic response.

Secondly, it is known that an endotoxin may lower serum calcium concentration by as much as 20 per cent; this could be due to calcium

TABLE I

EVIDENCE FOR THE DUAL ACTION OF PYROGEN ON THE CELLS OF THE ANTERIOR AND POSTERIOR HYPOTHALAMUS

Within the anterior hypothalamic "thermostat"

(1) Pyrogen injection evokes a fever
(2) Pyrogen suppresses electrical activity of warmth-sensitive neurons ("Noradrenoceptive" cells)
(3) Pyrogen increases electrical activity of cold-sensitive neurons ("Serotonoceptive" cells)
(4) Salicylate reinstates activity of pyrogen-depressed warmth-sensitive units
(5) Pyrogen releases 5-HT from platelets
(5-HT produces fever—noradrenaline lowers fever)
(6) Cooling elevates a pyrogen-produced hyperthermia to "hyperfever"
(7) Heating abolishes pyrogen fever

Within the posterior hypothalamic "set-point"

(1) Pyrogen fails to evoke fever
(2) Total ablation prevents fever from intravenous pyrogen
(3) Sodium ions elevate the "set-point"
(4) Calcium ions lower the "set-point"
(5) Intracerebral pyrogen causes a transient decrease in local ^{45}Ca levels

binding as a result of increases in plasma lactate, phosphate and lipoproteins, concentrations of all of which are raised after exposure to endotoxin (Skarnes, 1970). Since hyperthermia is produced by shifting the balance of sodium and calcium ions within the posterior hypothalamus in favour of sodium, it would appear that only a slight change in the ionic ratio of the serum flowing through the caudal hypothalamus is needed to cause the temperature rise (Myers and Veale, 1971). Patients suffering from chronic hypernatraemia or hypocalcaemia do not ordinarily exhibit the clinical signs of fever. Therefore, the alteration in the ionic ratio would be only transitory and facilitated probably by the pyrogen's action on the unknown interstitial system which maintains the ionic constituents at a constant level regardless of the fluctuations of serum ion concentration.

In a recent set of experiments Veale and I were surprised to find that in two of three cats ^{45}Ca decreased by 15 to 20 per cent in the perfusate collected from the posterior hypothalamus 30 to 90 minutes after *S. typhosa* was injected intravenously. During this time, ^{45}Ca concentrations in the mesencephalon and anterior hypothalamus were also changed (Fig.

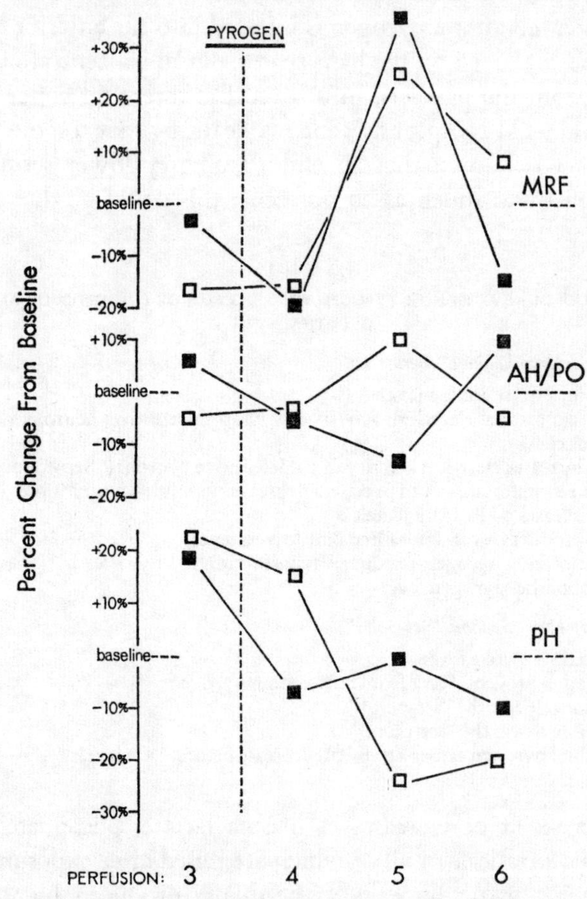

FIG. 7. Average percentage change (counts/min) of ^{45}Ca above its baseline in push-pull perfusates (5 min duration) in three brain-stem sites. *Top*: perfusions 3 to 6 of mesencephalic reticular formation (MRF) of three cats; *Middle*: perfusions 3 to 6 of anterior pre-optic region (AH/PO) of three cats; and *Bottom*: perfusions 3 to 6 of posterior hypothalamus (PH) of two cats. *S. typhosa* (0·2 ml of 1 : 5 dilution) was injected into the lateral cerebral ventricle 15 min after perfusion 3. ■—■: controls; □—□: pyrogen-treated cats. (From Veale, 1970.)

7). Another interesting point is that a salicylate which acts on the cells of the anterior hypothalamus (Wit and Wang, 1968) caused only a partial lowering of temperature (see Fig. 3). This could reflect a lack of effect of the drug on the posterior region in which the sodium-calcium ratio has been shifted and the set-point elevated.

In conclusion, a bacterial invasion results in a dual disturbance to the primate's normal hypothalamic function. Firstly, the circulating endotoxin increases the concentration of 5-HT released from the platelets in the anterior hypothalamus; secondly, and at the same time, the concentration of calcium in the posterior hypothalamus is temporarily lowered so that the sodium-calcium ratio is altered. Both of these changes could explain the great variety of intense physiological responses which occur during the progress of a pyrogen-induced fever.

SUMMARY

In the unanaesthetized rhesus monkey the cells of the anterior hypothalamus and pre-optic area are differentially sensitive to *Escherichia coli*, *Salmonella typhosa* and *Shigella dysenteriae*. Fevers are produced when these bacterial pyrogens are microinjected in regions along the mid-line.

The suggestion (Villablanca and Myers, 1965) that 5-hydroxytryptamine (5-HT) in the hypothalamus mediates an endotoxin fever appears to require modification since at least two other substances were released from the anterior hypothalamus of the monkey when this region was perfused during the course of an endotoxin fever. An increase in the concentrations of 5-HT was not consistently found in the hypothalamic perfusates.

From the experiments on the cat and monkey, a bacterial pyrogen seems to exert a dual action on the central nervous system. Firstly, in the anterior hypothalamus, 5-HT is liberated from platelets by a pyrogen in this region to produce hyperthermia; the thermosensitive neurons change their firing patterns at the same time. Secondly, in the posterior hypothalamus, in which the "set-point" for body temperature has been localized, pyrogen lowers the calcium concentration and because of the imbalance in the ratio of sodium to calcium ions an upward shift in the "set-point" occurs.

Acknowledgement

This research was supported in part by U.S.A. Office of Naval Research Contract N00014-67-A-0226-0003 and National Science Foundation Grant GB 7906. We are indebted to P. Curzon for his valuable technical assistance.

REFERENCES

ADLER, R. D., RAWLINS, M., ROSENDORFF, C., and CRANSTON, W. I. (1969). *Clin. Sci.*, **37**, 91–97.

ANDERSEN, H. T., HAMMEL, H. T., and HARDY, J. D. (1961). *Acta physiol. scand.*, **53**, 247–254.

ATKINS, E. (1960). *Physiol. Rev.*, **40**, 580–646.

BECKMAN, A. L., and EISENMAN, J. S. (1970). *Fedn Proc. Fedn Am. Socs exp. Biol.*, **29**, 523.
BENNETT, I. L., Jr, PETERSDORF, R. G., and KEENE, W. R. (1957). *Trans. Ass. Am. Physns*, **70**, 64–71.
BENZINGER, T. H. (1964). In *Homeostasis and Feedback Mechanisms*, pp. 49–80, ed. Hughes, G. M. London: Cambridge University Press.
COOPER, K. E. (1965). *Sci. Basis Med. A. Rev.*, 239–258.
COOPER, K. E., CRANSTON, W. I., and HONOUR, A. J. (1967). *J. Physiol., Lond.*, **191**, 325–337.
COOPER, K. E., GRUNDMAN, M. J., and HONOUR, A. J. (1968). *J. Physiol., Lond.*, **196**, 56–57P.
EISENMAN, J. S. (1969). *Am. J. Physiol.*, **216**, 330–334.
FELDBERG, W. S. (1971). This volume, pp. 115–124.
FELDBERG, W., and MYERS, R. D. (1965). *J. Physiol., Lond.*, **177**, 239–245.
FELDBERG, W., MYERS, R. D., and VEALE, W. L. (1970). *J. Physiol., Lond.*, **207**, 403–416.
FOX, R. H., and MACPHERSON, R. K. (1954). *J. Physiol., Lond.*, **125**, 21P
GANDER, G. W., CHAFFEE, J., and GOODALE, F. (1967). *Proc. Soc. exp. Biol. Med.*, **126**, 205–209.
GRANT, R., LEWIS, J., and AHRNE, I. (1955). *Fedn Proc. Fedn Am. Socs exp. Biol.*, **14**, 61.
HARDY, J. D. (1961). *Physiol. Rev.*, **41**, 521–606.
HAYWARD, J. N., and BAKER, M. A. (1968). *Am. J. Physiol.*, **215**, 389–403.
HUMPHREY, J. H., and JAQUES, R. (1955). *J. Physiol., Lond.*, **128**, 9–27.
IGGO, A. (1969). *J. Physiol., Lond.*, **200**, 403–430.
JACKSON, D. L. (1967). *J. Neurophysiol.*, **30**, 586–602.
MYERS, R. D. (1968). *Adv. Pharmac.*, **6**, 318–321.
MYERS, R. D. (1969). In *The Hypothalamus*, pp. 506–523, ed. Haymaker, W., Anderson, E., and Nauta, W. Springfield, Ill.: Thomas.
MYERS, R. D. (1970). *Physiol. Behav.*, **5**, 243–246.
MYERS, R. D., and BELESLIN, D. B. (1971). *Expl Brain Res.*, in press.
MYERS, R. D., KAWA, A., and BELESLIN, D. (1969). *Experientia*, **25**, 705–706.
MYERS, R. D., RUDY, T. A., and YAKSH, T. L. (1971). *Experientia*, in press.
MYERS, R. D., and VEALE, W. L. (1970). *Science*, **170**, 95–97.
MYERS, R. D., and VEALE, W. L. (1971). *J. Physiol., Lond.*, in press.
MYERS, R. D., and YAKSH, T. L. (1969). *J. Physiol., Lond.*, **202**, 483–500.
RANSON, S. W., CLARK, G., and MAGOUN, H. W. (1939). *J. Lab. clin. Med.*, **25**, 160–168.
RAŠKOVÁ, H., and VANĚČEK, J. (1964). *Pharmac. Rev.*, **16**, 1–45.
REPIN, I. S., and KRATSKIN, I. L. (1967). *Fiziol. Zh. SSSR*, **53**, 336–340.
ROSENDORFF, C., MOONEY, J. J., and LONG, C. N. H. (1970). *Fedn Proc. Fedn Am. Socs exp. Biol.*, **29**, 523.
SHEAGREN, J. N., WOLFF, S. M., and SHULMAN, N. R. (1967). *Am. J. Physiol.*, **212**, 884–890.
SHETH, V. K., and BORISON, H. L. (1960). *J. Pharmac. exp. Ther.*, **130**, 411–417.
SKARNES, R. C. (1970). *J. exp. Med.*, **132**, 300–316.
THOMPSON, R. H., HAMMEL, H. T., and HARDY, J. D. (1959). *Fedn Proc. Fedn Am. Socs exp. Biol.*, **18**, 159.
TULLY, J. G., GAINES, S., and TIGERTT, W. D. (1965). *J. infect. Dis.*, **113**, 445–455.
VEALE, W. L. (1970). Doctoral Dissertation, Purdue University.
VILLABLANCA, J., and MYERS, R. D. (1964). *Arch. Biol. Med. Exp., Chile*, **1**, 102.
VILLABLANCA, J., and MYERS, R. D. (1965). *Am. J. Physiol.*, **208**, 703–707.
WIT, A., and WANG, S. C. (1968). *Am. J. Physiol.*, **215**, 1160–1169.

DISCUSSION

Cooper: As far as I know no one has yet correlated these warm and cold firing units with actual thermoregulatory behaviour in the conscious intact animal. This makes it rather difficult to know for certain whether they are part of the system or not.

Myers: This is going to be a difficult technical problem to overcome because one can only lower these multi-barrel pipettes into deep structures of a large animal when it is anaesthetized and rigidly fixed in a head holder.

Pickering: But you can't investigate temperature regulation in an anaesthetized animal.

Myers: You can under urethane, and to some extent under pentobarbitone sodium. A certain amount of work has been done with the anaesthetized rabbit by Hardy and his group (Cabanac, Stolwijk and Hardy, 1968) and in the cat by Eisenman (1969). Furthermore, Wit and Wang (1968) found that in cats the discharge pattern of warmth-sensitive units was maintained when 0·8 to 1·2 g urethane/kg body weight were administered intraperitoneally. R. Lim (unpublished) also showed in the dog that a bacterial pyrogen evoked a fever when the animal was lightly anaesthetized with chloralose.

Pickering: If you change the blood temperature you can get vasoconstriction, vasodilatation and shivering. Does that happen with animals under urethane?

Myers: Yes, we have seen this in our experiments with cats (Myers R. D. and Veale, W., unpublished), and under pentobarbitone sodium anaesthesia Dr Feldberg and I (1964) observed shivering and vasoconstriction.

Pickering: Without any changes in sensitivity?

Myers: Hayward and Baker (1968) found that in the rhesus monkey, pentobarbitone sodium did affect the monkey's blood temperature in a warm environment, and this effect was dose dependent. The changes in vasoconstriction or vasodilatation under anaesthesia, however, could be controlled by artificial ventilation at a specified rate.

Rawlins: Eisenman (1969) and Wit and Wang (1968) demonstrated that warmth-sensitive cells in the hypothalamus decreased their sensitivities after intravenous injections of bacterial pyrogen, and Eisenman (1969) also showed that cold-sensitive cells increased their sensitivities. However, Eisenman (1969) found that the firing rates of these thermosensitive cells were unchanged at 38°C and therefore suggested that the normal drive for producing fever must come from another, as yet undetermined, class of cell.

Myers: You don't mean the sensitivity alone; you mean that the firing rate of the warmth-sensitive neurons decreased and that of the cold-sensitive neurons increased after intravenous Piromen (Eisenman 1969). The thermosensitive cells ("thermostat") are located predominantly in the anterior hypothalamus, and we believe the cells controlling the set-point are within the posterior hypothalamus (Myers and Veale, 1970). If this were not the case, one would have the uncomfortable difficulty of placing these two, perhaps independent, functional attributes of the diencephalon in one anatomical site. All the early evidence, which was based on lesion experiments and which dealt with the "heat maintenance" capacity of monkey, dog or cat (e.g. Ranson, Fisher and Ingram, 1937) placed this capacity in the posterior region of the hypothalamus. This classical work has been virtually overlooked in the last ten years.

Pickering: By a thermostat do you mean what I understand as a temperature receptor? Didn't you show that injecting endogenous pyrogen into the anterior hypothalamus produced fever? Isn't that really a changing of the set-point?

Myers: Hammel and co-workers (1963) have suggested that during a fever the set-point temperature could be shifted upward. In the region of the temperature-sensitive cells there are pyrogen-sensitive cells which may be one and the same. In response to pyrogen injected into the anterior hypothalamus the animals' set-point is elevated; this could be due to the efferent action of the anterior neurons on the posterior cells and possibly mediated by 5-HT cells in the anterior hypothalamus. Parenthetically, the newborn animal has no effective monoamine system (anterior hypothalamus) and yet it is born with a set temperature of 37°C (posterior hypothalamus) which appears to be ionically determined.

Pickering: If you put a pyrogen into the "thermostat" region the temperature may go up to 40°C, but the thermostat still receives the information and will regulate around it.

Cooper: If you microinject leucocyte pyrogen into the rabbit or cat brain, or if you inject bacteria into the monkey brain, you have to inject it into a fairly precise spot in the anterior hypothalamus. Therefore you would expect that something is happening to cells in that region. How then does leucocyte pyrogen affect the calcium concentration in the posterior hypothalamus? Are you postulating that you are getting some agent diffusing back to the posterior hypothalamic region which modifies the calcium content there?

Myers: Not at all. In my paper, we showed that during a "push-pull" perfusion 5-HT may be released from the cells of the anterior hypothalamus as a result of the pyrogen fever. There is a significant 5-HT release in about one out of seven cases; unfortunately, two other potent pharmacological substances are liberated as well. Although their nature is unknown, one appears to be a prostaglandin, and the other is a relaxing substance about which we have no information. So it appears that during the pyrogen fever three separate substances are released from the anterior pre-optic area alone; any one of these could be involved in transmitting the critical signal to the posterior hypothalamus in order to elevate the "set-point".

Work: By this use of whole bacterial cells as a pyrogen, I wonder whether you are not just getting an irritant action. Have you injected non-pyrogenic bacteria or have you removed all the pyrogen from cells?

Myers: No, we have not. However, we have gone to great pains to control for the possibility of a local irritating effect of a microinjection and have even studied the extent to which a fluid diffuses in brain tissue (Myers, 1966). The delay in a pyrexic response to a locally injected bacterial pyrogen would, I believe, preclude the possibility of some irritant action like cooling, heating or electrical stimulation which one would expect to exert a more immediate effect. Also, we have used three different kinds of pyrogen preparations, one of which was sonicated cells; thus, we were using disrupted cell walls only.

Work: The smallest molecular weight for a pyrogen that we know of is about 20 000. Could this size of molecule penetrate the blood-brain barrier and get into this region? It is known that endotoxins bind calcium and one would expect them to lower the calcium concentration if they do get in.

Cooper: I don't know whether bacterial pyrogens or lipolysaccharides can get in there. We put in some labelled lipolysaccharide intravenously and couldn't detect a significant amount in brain tissue which had had the blood removed (Cooper and Cranston, 1963). I don't know whether our specific activity was high enough to detect very small amounts that might be important at the sites of its action.

I am very interested in the experiment where you just put the bacterial pyrogen in and got fever, particularly in some of those experiments where you had such a short latency. We showed that in rabbits there was a difference in latency between the bacterial pyrogen and leucocyte pyrogen when put into this region (Cooper, Cranston and Honour, 1967) but possibly the bacterial pyrogen started to act when cells had been attracted to this region by its leucotaxic effect. Have you looked for leucocytes in sections of this region after you have put bacteria there, even after the short latency, and have you compared the intrahypothalamic sensitivity to leucocyte pyrogen and bacterial pyrogen in the monkey?

Myers: We haven't observed leucocytes under light microscopy at the site of injection; we usually look for glial cells, but we have not identified phagocytes. In the rhesus monkey, we have tried to culture leucocyte pyrogen by your method, Dr Cooper, but even by the intraperitoneal exudate procedure, we have been unsuccessful in doing so in three monkeys. The exudate, which should contain leucocytic pyrogen, had no effect when given intrahypothalamically.

Cranston: A prime-mover effect of calcium really hangs on the evidence in Fig. 7 (p. 144)—namely a fall of ^{45}Ca concentration in fluid removed from a "push-pull" cannula after injection of bacterial pyrogen. I find that very difficult to interpret because the posterior hypothalamus is the area in which Allen (1965) found evidence of uptake of radioactivity after intravenous injection of labelled plasma protein. What is the concentration of the ^{45}Ca in brain extracellular fluid in relation to that in the pyrogen? Could it be that the change only represents an increased permeability?

Myers: Only 50 to 200 microcuries of ^{45}Ca were given systemically an hour or so before we started the experiment (Fig. 7, p. 144). We found that the labelled calcium was distributed, as would be expected, in high concentrations throughout the body, but in very low concentration in the brain. I do not understand why ^{45}Ca concentrations in the anterior hypothalamus perfusates tend to rise, but fall in the posterior perfusates (Veale, 1970). This indicates to me that there is some sort of calcium selective mechanism functioning in an unknown way in the posterior region.

Cranston: But you are talking about the concentration in the "push-pull" fluid.

Myers: Yes. That is the concentration in extracellular fluid which is partially exchanged with the perfusion fluid at the local site.

Cranston: But it is going to be equilibration dependent. Presumably you are approaching a steady state of exchange of ^{45}Ca between circulation and brain extracellular fluid. If you assume that in this area you may now have an alteration in permeability, which would be compatible with these findings on labelled albumin, is it possible that what you are observing could be a consequence of this?

Pickering: What does a section of the bit of brain in which you have done your "push-pull" look like?

Cranston: The thermoregulatory system is intact after you have "push-pulled".

Myers: With our previous technique (Myers, 1967) there was some local damage; but now, by using the guide into which the "push-pull" cannulae are lowered only during a perfusion, the lesion is so small that we often have to put in dye to mark the site of perfusion for later histological analysis. If a slight expansion of tissue occurs during a perfusion of that site, the tissue must return to its original position shortly afterwards (see Myers, 1970).

Bondy: Are the data on Fig. 7 (p. 144) in absolute counts per minute?

Myers: No, in percentage deviation from baseline counts per minute.

Bondy: Are they higher or lower than plasma counts?

Myers: Plasma is much higher. From experiments using labelled magnesium and potassium it is known that the brain is relatively impermeable to ions in great excess.

Bondy: Can you explain this on the basis of the change in permeability? I don't see how you can: an increase in permeability would result in a rise in radioactivity.

Cranston: It is the non-protein-bound calcium counts in the plasma that one ought to be concerned about. Presumably they are much less than the total calcium counts in the plasma.

Myers: Certainly this is part of what we ought to be concerned about, because Skarnes (1970), using a calcium-selective electrode, quite clearly showed a text-book fall in serum calcium after the intravenous injection of a pyrogen.

TABLE I (Rawlins)

SERUM ELECTROLYTE CONCENTRATION AND ORAL TEMPERATURE IN HYPOPARATHYROIDISM

Patient	*Serum calcium* *mg/100 ml*	*Serum sodium* *mequiv/l*	*Oral temp.* °C
A	6·4	136	36·7
B	6·4	137	36·9
C	5·7	—	36·7
D	8·0	140	37·2
E	8·0	137	37·0
F	6·3	139	36·7
G	6·0	136	36·6

Normal ranges: serum calcium 8·7–10·0 mg/100 ml
serum sodium 134–143 mequiv/l

Bondy: You are interested in the ionized calcium, but that would represent probably only half the total plasma calcium. If the difference is much larger than that of the plasma calcium the argument would still hold. What are the ratios between what you are getting out of your cannulae and the blood?

Myers: The ^{45}Ca levels in the perfusates are only a tiny fraction of that found in an equivalent volume of blood.

Rawlins: Are you suggesting that the fall in serum calcium following intravenous injection of endotoxin is in part responsible for the fever?

Myers: Yes, but that is an oversimplification.

Rawlins: I have collected some data from seven patients with hypoparathyroidism. For each patient I have recorded the serum calcium, sodium and noon oral temperature on the day these chemical estimations were made (Table I). All these patients were normothermic and normonatraemic at a time when they were grossly hypocalcaemic.

Myers: I think the only explanation is that the change in the calcium concentration of the posterior hypothalamus might only be transitory.

From Skarnes' results, the serum calcium was lower only for a very short time after intravenous injection of pyrogen, and this parallels the results of our posterior hypothalamic perfusions in which we find that the change lasts about 90 minutes. This transitory change may nevertheless be sufficient to trigger the apparent shift in the set-point. If, as the data show, the pyrogen has a long-lasting action on the cells in the anterior region, then the pyrogen effect on these anterior cells could be precisely that which sustains the pyrogen fever once the set-point has been shifted by the imbalance in the calcium-sodium ratio in the posterior hypothalamus.

Landy: Dr Myers, your work with calcium and also the work of Skarnes seem to be exerted in the wrong direction. We (Rosen *et al.*, 1958) reported a lengthy series of experiments on a humoral system that inactivated endotoxin *in vitro*. Calcium played a profound role in these effects, but in the opposite direction. Unless normal serum calcium were tied up via EDTA, the inactivation of endotoxin by the serum endotoxin detoxifying component wasn't manifested.

Pickering: What effect of endotoxin were you measuring, Dr Landy?

Landy: Pyrogenicity for rabbits (Keene, Landy and Shear, 1961), among many other parameters; but there is evidence that the entire spectrum of activities that are characteristic of these components are progressively abolished. Long after we did this work there developed an awareness that this could also reflect a dissociation into subunits; it need not be destruction because in some cases these subunits can be reconstituted and again yield biological activity.

Myers: There may be some interdependence between the ions in the plasma and in the hypothalamus, but the stability of the concentration of ions in brain extracellular fluid in the face of severe alterations in serum concentration must be remembered. The mechanism of detoxification may simply be different in plasma and may be totally unrelated to ionic mechanisms which regulate the discharge patterns of the cells in the posterior hypothalamus.

Work: Skarnes (1968) has suggested that endotoxin produces its own inactivation by lowering the serum calcium and that esterases, with activity inversely proportional to the calcium concentration, attacked lipopolysaccharides. You may have esterases in the hypothalamus.

Bangham: Isn't the formation of a leucocyte pyrogen accepted as generally the final common path in the cause of fever in bacterial infections in all species? Are there other species in which a leucocyte pyrogen has not been found?

Myers: We have been trying to assay for monkey leucocyte pyrogen by microinjecting a preparation of monkey peritoneal exudate into the anterior hypothalamus of the monkey. A rise in temperature has not occurred.

Bodel: Bornstein has studied cats and dogs (Bornstein and Woods, 1969), we have worked with rabbits and humans, and Dr Atkins (unpublished) has studied guinea-pigs, and all these animals produce endogenous pyrogen.

Cranston: Can you give monkeys a fever with anything else? What happens if you give them *Staphylococcus*?

Myers: We have tried only *E. coli*, *Shigella dysenteriae* and *Salmonella typhosa*.

Cranston: Is it possible that these animals have become desensitized to bacterial pyrogen?

Myers: It is certainly possible, although fever can occur in this primate species. For example simian haemorrhagic fever nearly devastated a large part of the National Institutes of Health quarantine colony in 1964 (Palmer *et al.*, 1968). Perhaps monkeys desensitized to particular organisms are originating from one province of India.

Cooper: Have you spun off the cells of the vaccine that you inject into the brain and injected the supernatant?

Myers: Yes, with no effect.

Cooper: So it really is just a direct effect of the dead bacteria.

REFERENCES

ALLEN, I. V. (1965). *Br. J. exp. Path.*, **46**, 25.

BORNSTEIN, D. L., and WOODS, J. W. (1969). *J. exp. Med.*, **130**, 707.

CABANAC, M., STOLWIJK, J., and HARDY, J. (1968). *J. appl. Physiol.*, **24**, 645.

COOPER, K. E., and CRANSTON, W. I. (1963). *J. Physiol., Lond.*, **166**, 41–42.

COOPER, K. E., CRANSTON, W. I., and HONOUR, A. J. (1967). *J. Physiol., Lond.*, **191**, 325–337.

EISENMAN, J. (1969). *Am. J. Physiol.*, **216**, 330–335.

FELDBERG, W., and MYERS, R. D. (1964). *J. Physiol., Lond.*, **175**, 464–478.

HAMMEL, H., JACKSON, D., STOLWIJK, J., HARDY, J., and STROMME, S. (1963). *J. appl. Physiol.*, **18**, 1146.

HAYWARD, J. N., and BAKER, M. A. (1968). *Am. J. Physiol.*, **215**, 389–403.

KEENE, W. R., LANDY, M., and SHEAR, M. J. (1961). *J. clin. Invest.*, **40**, 302.

MYERS, R. D. (1966). *Physiol. Behav.*, **1**, 171.

MYERS, R. D. (1967). *Physiol. Behav.*, **2**, 373.

MYERS, R. D. (1970). *Physiol. Behav.*, **5**, 243.

MYERS, R. D., and VEALE, W. L. (1970). *Science*, **170**, 95–97.

PALMER, A., ALLEN, A., TAURASO, N., and SHELOKOV, A. (1968). *Am. J. trop. Med.*, **17**, 404.

RANSON, S. W., FISHER, C., and INGRAM, W. R. (1937). *Archs Neurol. Psychiat., Chicago*, **38**, 445.

ROSEN, F. S., SKARNES, R. C., LANDY, M., and SHEAR, M. J. (1958). *J. exp. Med.*, **108**, 701.

SKARNES, R. C. (1968). *J. Bact.*, **95**, 2031–2034.

SKARNES, R. C. (1970). *J. exp. Med.*, **132**, 300–316.

VEALE, W. L. (1970). Ph.D. Thesis, Purdue University.

WIT, A., and WANG, S. C. (1968). *Am. J. Physiol.*, **215**, 1160–1169.

RELEVANCE OF EXPERIMENTAL OBSERVATIONS TO PYREXIA IN CLINICAL SITUATIONS

W. I. CRANSTON, M. D. RAWLINS, R. H. LUFF and G. W. DUFF

Department of Medicine, St Thomas's Hospital Medical School, London S.E.1

PREVIOUS contributors to this symposium have considered the mechanism by which fever is caused in experimental situations, and the object of the present communication is to consider the ways in which fever may be caused in man and the extent to which observations in human disease are compatible with evidence obtained from animals.

Clearly, it is much simpler to design and execute experiments in animals, and for this reason most of our information comes from this approach. However, species variability makes it unjustifiable to extrapolate directly from the animal experiment to the situation in human disease, and thus it is important to attempt to determine whether fever in man behaves in the same way as fever in animals.

FEVER DUE TO INFLAMMATION

The most common cause of fever in man is the inflammatory process associated with infection, antigen-antibody reaction or necrosis. It is reasonable to attempt to explain this phenomenon on a hypothesis of a "final common path", and this is why the discovery of leucocyte pyrogen has provided a strong stimulus to research in this field.

We shall start with the premise that endogenous pyrogen may be released from leucocytes, monocytes, Kupffer cells, or other cells in the presence of inflammatory changes, whatever the origin of the inflammation. This begs the question of a possible direct action of bacterial pyrogen on the thermoregulatory apparatus, but if this is a cause of fever it is probably a very rare one.

There is evidence that human white blood cells produce leucocyte pyrogen which can be detected in man (Cranston *et al.*, 1956; Snell *et al.*, 1957) and in rabbits (Bodel and Atkins, 1966). We have confirmed that human leucocyte pyrogen can not only be detected but assayed quantitatively in rabbits; the assay does not appear to be influenced by the presence of human plasma in the injected material (Fig. 1).

In the rabbit, the production of pyrogen from leucocytes appears to be a two-stage process. There is an initial "activation" step, which involves the synthesis of protein within the cells (Moore *et al.*, 1970), followed by release of leucocyte pyrogen.

If we consider the ways in which fever might be caused during inflammatory processes it is apparent that a number of different mechanisms, all involving endogenous pyrogen, might be concerned.

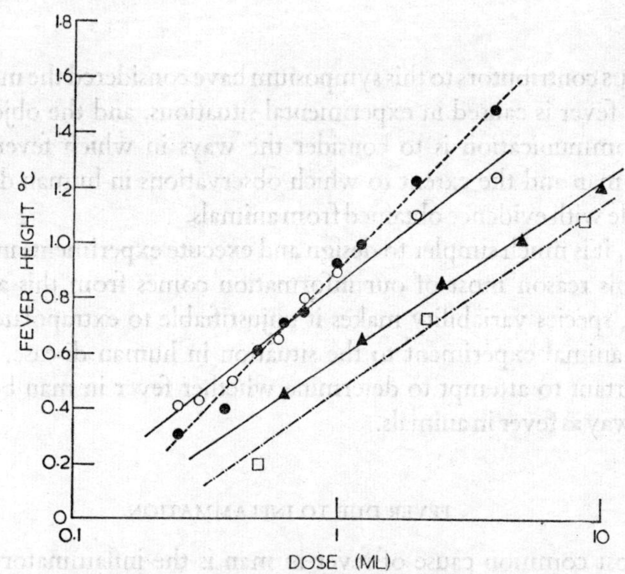

FIG. 1. Dose effect relationship for human endogenous pyrogen tested in groups of 6 rabbits. For comparison, dose effect curves from two assays of the effect of rabbit endogenous pyrogen (Murphy, 1966) are given. ●---●, ○——○: rabbit endogenous pyrogen in plasma; ▲——▲: human endogenous pyrogen in saline; □-·-·-□: human endogenous pyrogen in plasma.

Release of leucocyte pyrogen into the circulation

With a local lesion, such as an abscess, inflammatory cells are likely to be present in the affected area. These cells might produce endogenous pyrogen locally which is carried directly by the blood and/or lymph circulation to the central nervous system. If this is so, one might expect leucocyte pyrogen to be detectable in the inflammatory lesion itself, and possibly in the peripheral circulation if enough is present there. It might also reasonably be expected that the release of endogenous pyrogen should be slow, and therefore it might be difficult to detect in peripheral blood. This is certainly true in animals. Bennett (1956) was unable to detect endogenous pyrogen in the peripheral blood of animals with pneumococcal peritonitis, using up to 80 ml of whole blood or serum, though endogenous pyrogen was present in the peritoneal exudate and in the lymph. Subsequently, however, King and Wood (1958) were able to demonstrate leucocyte pyrogen in the peripheral blood after similar infections, using 50–100 ml of serum. These are large amounts relative to the rabbit's circulating volume, and if the same quantitative relationship holds in man it is not surprising that attempts to demonstrate circulating endogenous pyrogen have largely failed in human disease (Snell, 1961), despite the presence of a material with the properties of endogenous pyrogen in inflammatory lesions or tumours associated with fever (Snell, 1962; Rawlins, Luff and Cranston, 1970). Endogenous pyrogen has also been detected in the joint fluid of patients with non-bacterial arthritis, though the presence of this pyrogen appeared to be unrelated to the nature of the joint disease, and presumably also to the patient's temperature at the time of aspiration of fluid (Bodel and Hollingsworth, 1968). In two out of four patients with malaria a pyrogen with properties of endogenous pyrogen was found in peripheral blood taken while the patient's temperature was rising rapidly (Cranston, 1966); no pyrogen was detected in blood taken while the patients were afebrile. All this evidence might be compatible with the hypothesis that endogenous pyrogen is present in peripheral blood of patients with fever, but the concentration is so low that it cannot be detected except where there is a rapid rise of temperature and presumably a relatively high concentration of endogenous pyrogen in the circulation. Thus, in most cases, the situation might be similar to that prevailing when a slow continuous infusion of leucocyte pyrogen is given.

This hypothesis can be tested by trying to detect leucocyte pyrogen in the circulation during fever induced by a continuous infusion of this agent. In rabbits, leucocyte pyrogen was prepared by incubating whole blood (100 ml) with "E" pyrogen (Organon Laboratories Ltd., Morden, Surrey, U.K.) (0·3 μg) for 18 hours and aspirating the plasma. The plasma was then injected into donor animals in a dose of 2·5 ml followed by a continuous infusion of leucocyte pyrogen at the rate of 0·02 ml/min. This resulted in a sustained temperature rise of about 1·5°C which was maintained for the duration of the infusion. One, two, and four hours after the start of the infusion animals were exsanguinated by cardiac puncture and the blood centrifuged. Pooled plasma in a dose of 20 ml was injected into eight recipients, and it is clear from Fig. 2 that a considerable amount of leucocyte pyrogen was present at all times, though the amount detected did show a significant decrease between one and four hours after the start of the infusion.

Similar experiments have been carried out in four normal human subjects. Leucocyte pyrogen was prepared, as before, by incubating heparinized whole blood with "E" pyrogen. The plasma was then injected into a peripheral vein in a dose of 25 ml followed by a continuous infusion of leucocyte pyrogen at a rate of 0·1 ml/min. As in the rabbit, this resulted

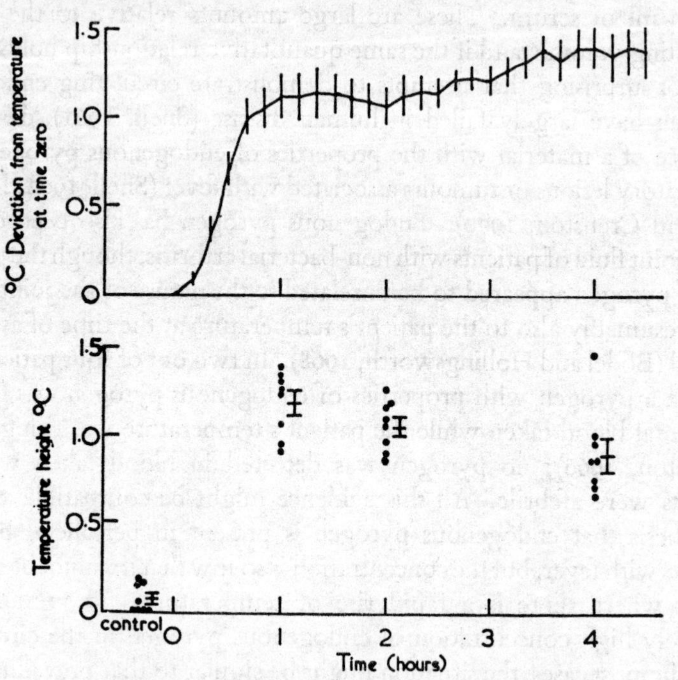

FIG. 2. Upper part—mean temperature change in rabbits injected with 2·5 ml of rabbit endogenous pyrogen, followed by a continuous infusion of leucocyte pyrogen at 0·02ml/min.
Lower part—transferable endogenous pyrogen in plasma samples taken 1, 2 and 4 hours after the injection of endogenous pyrogen. Vertical bars indicate ± 1 S.E.M.

in a sustained elevation of temperature of about 1–1·5°C (Fig. 3). Two and a half hours after the start of the infusion about 500 ml of blood was withdrawn from a vein in the opposite arm; acid citrate-dextrose was used as the anticoagulant, and in these experiments the blood was immediately centrifuged; in three of the subjects plasma and cells were separated and reinjected separately on subsequent days. None of these reinjections was followed by any temperature rise or subjective manifestations. In the fourth experiment, injection of 20 ml of plasma into each of six rabbits caused no fever. It is therefore clear that in man a considerable rise in body temperature can be present due to the intravenous infusion of leucocyte pyrogen at a time when no pyrogen can be detected in the circulation, and it would appear to follow that the search for transferable pyrogen in blood is unlikely to be a fruitful pursuit in human fever.

Leucocyte pyrogen is probably a relatively small molecule, and the possibility exists that it might be filtered at the renal glomerulus. Sokal and

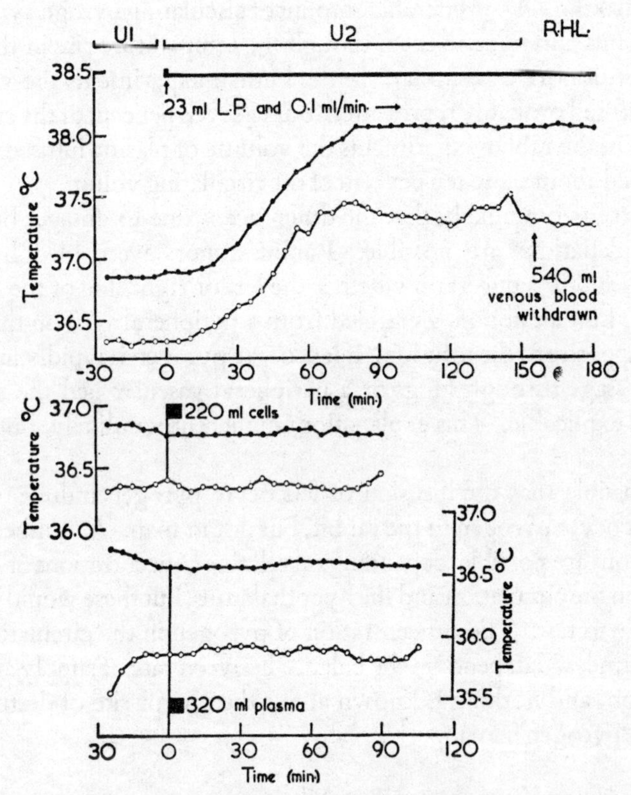

FIG. 3. Upper part—response of one normal subject to injection and infusion of autologous leucocyte pyrogen.
Lower part—temperature response of the same subject to reinfusion of his cells and plasma given separately. ●——●: rectal temperature; ○——○: ear temperature.

Shimaoka (1967) detected pyrogenic material in the urine of febrile patients with Hodgkin's disease, and this material resembled endogenous pyrogen in its properties. An attempt was made to detect pyrogen in the urine of our subjects who received leucocyte pyrogen. Urine samples were concentrated by dialysis against polyethylene glycol and injected into groups of rabbits in volumes equivalent to the original volume of urine that contained 10 mg creatinine (that is roughly 0·1 per cent of 24-hour urine volume). Neither control urine samples, nor those taken during fever induced by pyrogen infusion, were pyrogenic; this, however, does not invalidate the possible use of urine assays for leucocyte pyrogen, because there is some evidence of loss of leucocyte pyrogen in the concentration process and it will be necessary to pursue this line further.

The difference between rabbit and man in these infusion experiments is rather striking. A considerable amount of circulating pyrogen was present in the rabbits and none in man, though the temperature rise in the human and rabbit donors was similar. In the human experiments the volume of plasma infused probably represented four to seven per cent of the circulating volume. In the rabbit experiments the volume of plasma infused probably represented about six to ten per cent of the circulating volume.

It is extremely unlikely that the difference is due to dosage, but several other explanations are possible. Rabbit donors were bled by cardiac puncture, and it is uncertain whether the left or right side of the heart was aspirated; human donors were bled from a peripheral vein on the arm on the side opposite to the infusion. If leucocyte pyrogen is rapidly inactivated in one passage through lungs or a peripheral vascular bed the difference might be explicable. This explanation is inherently unlikely, but it can be tested.

It is possible that the infusion of leucocyte pyrogen induces release of more leucocyte pyrogen in the rabbit, but not in man. A number of other explanations are possible, depending on relative concentrations of leucocyte pyrogen in the circulation and the hypothalamus, but these would be almost impossible to test. The concentration of pyrogen in the circulation at any point in time will depend on the balance between rate of supply and rate of destruction, and nothing is known about the rate or site of destruction of leucocyte pyrogen in man.

Release of activated leucocyte into the circulation

Another possible way in which a local lesion might induce fever is by an escape of activated cells from the lesion into the circulation. These cells might remain in the circulation, releasing pyrogen, or they might be trapped in the lungs or reticuloendothelial system, releasing leucocyte pyrogen from that site. This is a very speculative proposition with little evidence to confirm or refute it. Cells from peritoneal exudates, which are presumably "activated", do not cause fever when injected intravenously into rabbits (Fessler *et al.*, 1961); this is not surprising in view of the effect of potassium in suppressing release of leucocyte pyrogen from activated cells in peritoneal exudates (Berlin and Wood, 1964). Leucocytes obtained from inflammatory human joint fluid did not release pyrogen when incubated with saline, though they could release rather small amounts of pyrogen when activation was induced by phagocytosis (Bodel and Hollingsworth, 1968). The presence of pyrogen in the cell-free joint fluid suggested that the cells had previously released leucocyte pyrogen, and that this was why the amount of pyrogen released after phagocytosis was small. The implication is that these cells, previously "activated", had had this activation suppressed. An alternative possibility is that the pyrogen detected in joint fluid supernatants was produced by cells other than those found in the joint. We have failed to find evidence of activated white blood cells in the peripheral blood of two febrile patients with renal carcinoma whose tumours contained pyrogen (Rawlins, Luff and Cranston, 1970). However, the site of sampling could be important in this kind of investigation. If activated cells were, for example, sequestered in the lungs, they might only be detectable in blood sampled from the venous drainage pathway of the inflammatory lesion. There is as yet no evidence that cells activated by bacterial pyrogen or by phagocytosis can cause fever if they are injected without preliminary incubation. This is clearly a field in which further study would be worth while.

Release of bacterial pyrogen into the circulation

The role of bacterial endotoxin in human infections caused by gram-negative organisms had been clarified by the work of Greisman and his colleagues (Greisman, Young and Woodward, 1966; Greisman *et al.*, 1967). These authors demonstrated that tolerance to intravenous injections of endotoxin could be developed during the progress of typhoid fever in man; the induction of tolerance to typhoid endotoxin before the development of typhoid fever resulted in continued protection against the pyrogenic action of injections given during the illness. Without such prior induction of tolerance, sensitivity to the pyrogenic action of intravenous injections of homologous or heterologous endotoxin was greatly enhanced during typhoid fever; this sensitivity did not appear to be due to a general depression of the reticuloendothelial function. Induction of tolerance to typhoid endotoxin before the development of typhoid fever did not appear to influence the febrile course of the disease.

Continuous infusion of typhoid endotoxin in normal subjects and in patients with typhoid fever caused an initial temperature rise, which subsided over a few hours, while the infusion continued. The conclusion must be that circulating endotoxin plays a negligible part in the production of fever in this infection. The possibility is not excluded that bacterial endotoxin may have a local effect in certain tissues, though this again is a hypothesis which can be tested.

FEVER DUE TO PYROGENIC STEROIDS

The interaction between 5β steroids and human white blood cells has already been considered in some detail at this meeting. It is clear that C_{19}, C_{21} and C_{24} steroids with a 5β configuration can cause a marked pyrexia when injected intramuscularly in man, or when given as a slow intravenous infusion, though they have much less effect after rapid intravenous injection (Kappas *et al.*, 1959; Kappas, Glickman and Palmer, 1960). Similar injections into animals of many different species have failed to cause any febrile response (Kappas, Glickman and Palmer, 1960). The prolonged incubation of one of these steroids, aetiocholanolone, with human white blood cells results in the release of leucocyte pyrogen (Bodel and Dillard, 1968), but its incubation with rabbit cells has no such effect.

This represents an undoubted species difference, but it is very difficult to decide whether the phenomenon is of clinical importance. Early reports (Bondy *et al.*, 1958; Huhnstock, Kuhn and Oertel, 1966) suggested that there might be more free aetiocholanolone in the plasma of some patients during attacks of febrile illness of unknown cause. This, however, has not been confirmed by George and co-workers (1969) using a much more sensitive method than the earlier workers for measuring free aetiocholanolone. They found no difference in the amount of free aetiocholanolone in plasma during febrile or afebrile periods in 20 patients with familial Mediterranean fever and 20 patients with recurrent undiagnosed febrile illness. In both groups of patients, however, the plasma levels of free aetiocholanolone were slightly, but significantly, higher than those in normal control subjects. The meaning of this is uncertain, but these observations must cast considerable doubt on the role of aetiocholanolone in febrile illness. This leaves the possibility that breakdown products of progesterone, pregnenolone or pregnanediol might be related to the temperature changes occurring with ovulation. Both of these steroids cause fever in man after intramuscular injection (Kappas *et al.*, 1959).

Since there is evidence that oestradiol can suppress the release of leucocyte pyrogen from human white blood cells incubated with aetiocholanolone (Bodel *et al.*, 1969), this would form a very tidy hypothesis. If this is the only mechanism responsible for temperature changes with ovulation, then the rider should be that ovulation in species insensitive to 5β steroids should not be accompanied by any marked change of temperature. We have been unable to find out whether this is generally true.

MALIGNANT HYPERPYREXIA DURING ANAESTHESIA

This syndrome was brought to general notice by Denborough and Lovell (1960) when they reported the survival of a patient who became comatose and very hot after the administration of a general anaesthetic for a minor procedure. This patient knew that other members of his family had died during anaesthesia. Since then a large number of similar cases have been reported. There has been no clear indication that any single anaesthetic agent is related to the development of the condition, though halogenated anaesthetics and suxamethonium chloride have been implicated in most of the recorded cases; the condition appears to be inherited as an autosomal dominant.

During anaesthesia in such cases there is increasing muscle stiffness and difficulty in inflating the lungs. Thereafter, progressive muscle spasm, hyperpyrexia and coma ensue, and the death rate from this condition has exceeded 70 per cent (Isaacs and Barlow, 1970). There is increasing evidence that this condition is associated with a clinical or subclinical myopathy. Relatives of patients who have died of this condition have shown high levels of serum creatine phosphokinase, not necessarily associated with overt clinical myopathy (Isaacs and Barlow, 1970; Denborough *et al.*, 1970*a*). During the hyperpyrexia hypocalcaemia has been observed, probably due to a rise in plasma phosphate caused by muscle damage. Very high levels of creatine phosphokinase have been observed during this hyperpyrexia (Denborough *et al.*, 1970*b*). A similar condition has been reported in pigs (Hall *et al.*, 1966; Harrison *et al.*, 1968) where suxamethonium chloride was considered to be the precipitating agent. High levels of creatine phosphokinase have been observed in these animals also.

The available evidence suggests that this condition is primarily an abnormality of muscle, though there is not yet convincing evidence that other tissues do not participate in increased heat production. Berman and co-workers (1969) have suggested, on the basis of measurements of oxygen consumption in pigs, that much of the increased heat production comes from anaerobic sources, but their evidence is not set out sufficiently fully for this to be assessed. They also observed an early increase in liver temperature during anaesthesia; but again inadequate evidence is provided to show whether this indicates thermogenesis by tissues other than muscle.

SUMMARY

The evidence for the participation of endogenous pyrogen in human disease is considered. There is evidence that leucocyte pyrogen can be found in inflammatory exudates and in some tumours, but attempts to detect this material in the circulation of febrile patients have been generally unsuccessful. This does not mean that no endogenous pyrogen is present as it is impossible to detect this material in the blood of normal subjects made febrile by a continuous infusion of endogenous pyrogen.

The possibility has been considered that "activated" circulating white blood cells might be involved in human fever, but there is inadequate evidence to support or deny this proposition.

The possible contribution of 5β steroids to clinical fever has been considered.

Acknowledgements

R. H. Luff is supported by a grant from the National Fund for research into crippling diseases, and we are also grateful to the Medical Research Council for the loan of equipment.

REFERENCES

BENNETT, I. L. (1956). *Bull. Johns Hopkins Hosp.*, **98,** 216–235.
BERLIN, R. D., and WOOD, W. B. (1964). *J. exp. Med.*, **119,** 697.
BERMAN, M. C., HARRISON, G. G., DU TOIT, P., BULL, A. B., and KENCH, J. E. (1969). *S. Afr. med. J.*, **43,** 545–546.
BODEL, P., and ATKINS, E. (1966). *Proc. Soc. exp. Biol. Med.*, **121,** 943–946.
BODEL, P., and DILLARD, M. (1968). *J. clin. Invest.*, **47,** 107–117.
BODEL, P., DILLARD, G. M., KAPLAN, S., and MALAWISTA, S. E. (1969). *J. clin. Invest.*, **48,** 9a–10a.
BODEL, P. T., and HOLLINGSWORTH, J. W. (1968). *Br. J. exp. Path.*, **49,** 11–19.
BONDY, P. K., COHN, G. L., HERRMAN, W., and CRISPELL, K. R. (1958). *Yale J. Biol. Med.*, **30,** 395–405.
CRANSTON, W. I. (1966). *Br. med. J.*, **2,** 69–75.
CRANSTON, W. I., GOODALE, F., SNELL, E. S., and WENDT, F. (1956). *Clin. Soc.*, **16,** 615–626.
DENBOROUGH, M. A., EBELING, P., KING, J. O., and ZAPF, P. (1970*a*). *Lancet*, **1,** 1138–1140.
DENBOROUGH, M. A., FORSTER, J. F. A., HUDSON, M. C., CARTER, N. G., and ZAPF, P. (1970*b*). *Lancet*, **1,** 1137–1138.
DENBOROUGH, M. A., and LOVELL, R. R. H. (1960). *Lancet*, **2,** 45.
FESSLER, J. H., COOPER, K. E., CRANSTON, W. I., and VOLLUM, R. L. (1961). *J. exp. Med.*, **113,** 1127.
GEORGE, J. M., WOLFF, S. M., DILLER, E., and BARTTER, F. C. (1969). *J. clin. Invest.*, **48,** 558–563.
GREISMAN, S. E., HORNICK, R. B., WAGNER, H. N., and WOODWARD, T. E. (1967). *Trans. Ass. Am. Physns*, **80,** 250–258.
GREISMAN, S. E., YOUNG, E. J., and WOODWARD, T. E. (1966). *Clin. Res.*, **14,** 332.
HALL, L. W., WOOLF, N., BRADLEY, J. W. P., and JOLLY, D. W. (1966). *Br. med. J.*, **2,** 1305.
HARRISON, G. G., BIEBUYCK, J. F., TERBLANCHE, J., DENT, D. M., HICKMAN, R., and SAUNDERS, S. J. (1968). *Br. med. J.*, **3,** 594–595.
HUHNSTOCK, K., KUHN, D., and OERTEL, G. W. (1966). *Dt. med. Wschr.*, **91,** 1641.
ISAACS, H., and BARLOW, M. B. (1970). *Br. Med. J.*, **1,** 275–277.
KAPPAS, A., GLICKMAN, P. B., and PALMER, R. H. (1960). *Trans. Ass. Am. Physns*, **73,** 176–185.
KAPPAS, A., SOYBEL, W., FUKUSHIMA, D. K., and GALLAGHER, T. F. (1959). *Trans. Ass. Am. Physns*, **72,** 54–61.

KING, M. K., and WOOD, W. B. (1958). *J. exp. Med.*, **107**, 305–318.
MOORE, D. M., CHEUK, S. F., MORTON, J. D., BERLIN, R. D., and WOOD, W. B. (1970). *J. exp. Med.*, **131**, 179–188.
MURPHY, P. A. (1966). D.Phil. thesis, University of Oxford.
RAWLINS, M. D., LUFF, R. H., and CRANSTON, W. I. (1970). *Lancet*, **1**, 1371–1373.
SNELL, E. S. (1961). *Clin. Sci.*, **21**, 115–124.
SNELL, E. S. (1962). *Clin. Sci.*, **23**, 141–150.
SNELL, E. S., GOODALE, F., WENDT, F., and CRANSTON, W. I. (1957). *Clin. Sci.*, **16**, 615–626.
SOKAL, J. E., and SHIMAOKA, K. (1967). *Nature, Lond.*, **215**, 1183–1185.

DISCUSSION

Pickering: Were the cells you injected into the malarial patient suspended in saline?

Cranston: No, we gave them packed cells.

Pickering: How do you get them in?

Cranston: By gravity.

Pickering: Could those leucocytes have been releasing leucocyte pyrogen?

Cranston: I can't prove that they weren't releasing leucocyte pyrogen.

Pickering: Has anybody tried injecting a rabbit with a huge dose of bacterial pyrogen and when the fever was high taken out blood and separated it, and then injected the cells and plasma?

Bodel: We have not injected the cells directly but we have incubated the cells, and they do not spontaneously release pyrogen, although in rabbits the plasma drawn at the same time contains endogenous pyrogen. We have not found circulating cells to be producing pyrogen either in patients with fever or in rabbits with artificially-induced fevers (Atkins E., and Bodel, P. T., unpublished). Dr Cranston, how do you explain the fever that you got with the cell fraction from the malarial patient?

Cranston: On the basis of the time of onset it looked as though the cell fraction contained leucocyte pyrogen.

Snell: The onset was very slow.

Cranston: The temperature began to rise in under 40 minutes.

Snell: This latency is dependent on dose as well as type of pyrogen. With leucocyte pyrogen, fever as high as that should have come on a bit earlier.

Atkins and I (Snell and Atkins, 1967) added bacterial pyrogen to rabbit whole blood *in vitro*, then rapidly centrifuged the blood in the cold and took off the plasma. There was then little bacterial pyrogen in the plasma, but it was mostly in the cells, for when the cells were injected they gave a bacterial pyrogen type of fever. When aliquots of these cells were incubated, the bacterial pyrogen disappeared and leucocyte pyrogen appeared both in the plasma and in the cells.

Cooper: In the malarial situation malarial parasites must also have been present in the cell fraction and they may have had some antigenic behaviour.

Cranston: We had controls in both these patients; we injected blood taken when they were not having their febrile episodes but while they had parasites in their red cells.

Myers: The fact that you withdrew the blood on a rising phase in the plasmodium cycle poses a problem with regard to refractoriness or tachyphylaxis. If there is such a thing as tachyphylaxis to a pyrogen, what do you think is the mechanism whereby the temperature of a patient can cycle, say, every other day?

Cranston: This might be compatible with release of pyrogen from peripheral leucocytes.

Myers: Then that would serve as the proof that this particular pyrexic response would be due to leucocyte pyrogen.

Cranston: It would be nice to have more definite evidence that this was so.

Work: Not all non-lipopolysaccharide pyrogens produce refractoriness; does this malarial one?

Cranston: I don't know.

Pickering: Dr Cranston, your cell fractions contain red cells, white cells, red cell envelopes and parasites, and it might be the parasites which are pyrogenic. A mosquito bite doesn't produce fever, but if lots of bits of parasite were injected into the blood-stream, would the temperature go up?

Cranston: I'm not sure how you would get these bits of parasite.

Rawlins: You could culture them.

Landy: If one used the model Dr Atkins talked about one could harvest peripheral blood cells from the sensitized subject, expose them to antigen in short-term tissue culture, a process which itself is non-pyrogenic, and then reinfuse them into the same patient.

Cranston: Or it could be done on a rabbit.

Snell: You suggested that the rate of clearance of circulating pyrogen might be different in man and in the rabbit. How long does the fever last when you stop the intravenous infusion? Have you done this on man?

Cranston: Yes, but we haven't gone on until the temperature goes right down. If there is a relationship between the hypothalamic concentration of leucocyte pyrogen and the temperature response, temperature changes would not tell us about plasma concentrations, but about concentrations in the hypothalamus, or at any other site at which pyrogen acts. The temperature stays up for over an hour in man when you stop the infusion.

Rawlins: In rabbits with a steady-state fever induced with an intravenous priming injection, followed by a sustaining infusion of homologous leucocyte pyrogen, little change in temperature occurred for about an hour after stopping the infusion. Similarly, in human volunteers with a steady-state fever induced by an intravenous infusion of leucocyte pyrogen, there is no change in temperature for at least 50 minutes after stopping the infusion.

Cranston: It is possible to make a rough estimate of the rate of clearance of leucocyte pyrogen after intravenous injection, from the data of Bornstein, Bredenberg and Wood (1963). If it is assumed that the clearance of leucocyte pyrogen is exponential, and that the fever index is proportional to the logarithm of the dose of leucocyte pyrogen in the range in which these authors were working, the half time is almost exactly 10 minutes. This would indicate a rapid removal of leucocyte pyrogen, with about 7 per cent of the material in the circulation being removed each minute.

Bodel: Bornstein, Bredenberg and Wood (1963) also showed that 30 minutes after injection of a large dose of leucocyte pyrogen, no transferable pyrogen could be obtained from the plasma. Atkins and Huang (1958) got essentially similar data using virus-induced endogenous pyrogen.

Snell: Then if Atkins gave an enormous dose he got a second wave of circulating pyrogen, as if the injected leucocyte pyrogen acted as a stimulus itself to release more pyrogen. This might happen in the rabbit and not in man.

Cranston: I'm not too happy about explaining the difference between the species by saying that in one a given dose of leucocyte pyrogen will produce an effect, but in another it won't. We gave relatively small priming doses, nothing like the size of the dose that Atkins gave.

Whittet: I found bacterial pyrogen disappearing and then coming back again in virgin rabbits (Whittet, 1964). Atkins and Wood (1955) said that they only got this in sensitized rabbits. I gave a dose of pyrogen, took blood out at various intervals and tested it in other rabbits. I got a curve coming down almost to zero and then a rise.

Snell: The second rise was the leucocyte pyrogen. The rate of clearance of bacterial pyrogen in the rabbit was found by Atkins to be very rapid if the animal had previously experienced bacterial pyrogen (sensitized), and it could not be detected in the same form later.

Whittet: My curves with several unsensitized rabbits were the same as those of Atkins and Wood (1955) with sensitized rabbits.

Bondy: It has been suggested that the leucocyte pyrogen may be cleared somewhere between the site of injection and the peripheral vein which you sample. Have you thought about using the catheter techniques to drain proximal to your infections?

Cranston: If we were looking for clearance of circulating pyrogen in the lung we would have to do arteriovenous differences which would mean a lot of assay problems. A simpler way to solve this might be to give continuous infusion separately to each side of the lung.

Bondy: In man you don't have that problem because you are getting nothing from the peripheral venous circulation, so that if you find anything on the other side you would be subtracting zero from it.

Cranston: Are you suggesting that we infuse pyrogen into the arm and sample from the right atrium?

Bondy: Yes, or if you wanted to do it with a naturally occurring disease, for example to pick up endogenous pyrogen from a liver abscess, you could put a catheter in the right atrium.

Cranston: I would much rather do it in someone whom we knew that we had infused leucocyte pyrogen. If we then cannot detect it, the problem must be a quantitative one.

Bondy: That is a little like rigging it.

Cranston: Yes, but there may still not be a high enough concentration with that infusion rate to find pyrogen in a sample of practicable size.

Bondy: But what happens to it in that 6 or 8 inches of vein?

Cranston: I'm not sure about that. If we infuse leucocyte pyrogen into one arm, at 0·1 ml/min, it goes somewhere and none of it is detectable in venous blood from the opposite arm. If we now aspirate from the same side as the infusion, we ought to have a small quantity of pyrogen in our aspirate, but there might still not be a detectable amount.

Bondy: Could the leucocyte pyrogen be picked up by a cell and changed into something which you can't shake loose?

Cranston: No, the problem might only be a quantitative one; if we bleed at 540 ml/6 or 7 min while we infuse leucocyte pyrogen into the same arm at 0·1 ml/min, we cannot collect more than 0·6 or 0·7 ml of leucocyte pyrogen, even assuming that our aspiration system removes all the infused pyrogen. This is a small quantity of leucocyte pyrogen, and I do not know whether it would be detectable.

Bondy: You may not need to take so much as you have essentially a closed system.

Cranston: We know we can't have more than 0·6 or 0·7 ml, plus any undetectable quantity supplied in arterial blood.

Cooper: The pyrogenic situation occurring in tissue rejection after an organ transplant is one which could involve leucocyte and tissue pyrogens.

Cranston: It would have to be an enormous rejection response for there to be any chance of finding circulating pyrogens. We could not detect circulating pyrogen, despite a rise in temperature of 1·5°C produced by the leucocyte pyrogen that we infused.

Bangham: Many grafts are made under the influence of immunosuppressive drugs. At what stage in their development are the white cells involved? Does a patient with aplastic anaemia get fever in response to an infection?

Cranston: Yes.

Bangham: And do patients with any simple leucocytopenia?

Cranston: Yes, but they may have white cells in other tissues.

Pickering: Bennett and Beeson (1953) in their original experiment showed that if the granulocytes were suppressed by nitrogen mustard the patients still got fever.

Bodel: There is evidence (Herion, Walker and Palmer, 1961; Gillman, Bornstein and Wood, 1961) that nitrogen mustard treatment reduces febrile responses of rabbits to bacterial pyrogen (endotoxin). However, we now know that the polymorphonuclear leucocyte is not theoretically necessary for fever since mononuclear cells are very effective pyrogen producers. It is also true that nitrogen mustard-treated rabbits, like agranulocytic patients, have high fevers.

Landy: The complex graft rejection studies involving pyrogenic effects could be approached directly by using the *in vitro* model of the homograft reaction; this involves mixed lymphocyte cultures between peripheral blood cells of histo-incompatible donor and recipient. The product of this reaction might very well be pyrogenic.

Feldberg: In the rare cases of severe and even lethal pyrexia following anaesthesia, what is the evidence, Professor Cranston, that the muscular contractions associated with the pyrexia are peripheral effects?

Cranston: The evidence is fairly good. There is a considerable increase in peripheral muscle tension in man and animals. In animals, transection of the nerve to the muscle has no effect once the contraction has started. In man, other muscle relaxants do not affect the contracture. The patients appear to be peripherally vasodilated. This is not like the pyrogenic situation in which you nearly always get a peripheral vasoconstriction during the rise in temperature. Oxygen consumption is enormously increased. There is a lot of evidence of peripheral abnormality. Some of these patients, and some of the families of these patients, have evidence of myopathy and high levels of creatine phosphokinase and aldolase in their peripheral blood. Their serum calcium drops when they get hyperpyrexia, but this is almost certainly due to phosphate being released from the injured muscle.

Whittet: That is more like nitrophenol.

Cranston: Yes. I don't know whether any cells other than the muscle cells participate in this. Berman and co-workers (1969) found that liver temperature goes up early, but as they don't appear to have measured muscle temperature, you can't tell whether that is cart or horse. They also claim that a lot of the increased energy production must be anaerobic on the basis of oxygen consumption measurements.

Bondy: Are you implying that members of the family who don't have this disease have high plasma enzyme levels?

Cranston: Yes, and there is evidence that in some of these groups it is inherited.

Pickering: People with a cerebral haemorrhage get a higher fever when it bursts in the ventricles.

Cranston: I have always taken the rather simple view that one of the mechanisms is merely interference with efferent pathways.

Pickering: Wouldn't that make them hypothermic?

Cranston: It depends on which pathways are affected. Professor Feldberg might suggest that they were releasing monoamines.

Feldberg: It is possible that monoamines may be released as a result of a spasm of the perforating vessels. Myers and I have discussed this possibility for the pyrexia which occurs in certain cases of brain injury following car accidents. When we perfused the cerebral ventricles of anaesthetized cats the effluent contained small amounts of 5-HT, and when the cat was killed and the perfusion was continued, the 5-HT output increased greatly for an hour or two. If a similar condition of insufficient circulation was produced in some parts of the hypothalamic region by the spasm of the perforating vessels it might lead to an increased release of 5-HT which, in turn, might produce the fever.

Myers: We have done experiments in which, by accident, a bloody perfusate is collected from the anterior pre-optic area. In these cases, the animal develops a raging fever similar in many ways to the kind of hyperthermia you see when 5-HT is microinjected into the same region. Also, when blood is carried into the cerebrospinal fluid as a result of damage caused by an injector needle passing through tissue, a fever occurs which again is similar to that produced by 5-HT.

Cooper: If you did this in the rabbit you ought to get hypothermia due to the released 5-HT, if this is the explanation.

Cranston: If you put 0·25 ml heparinized blood into the cisterna magna of the rabbit it is not pyrogenic, but the situation may be different if one uses blood without anti-coagulant; we have not examined this.

Feldberg: Substances injected into the cisterna magna can reach the anterior hypothalamus from the outer surface of the ventricular wall because the cerebrospinal fluid passes from the cisterna to the ventral surface of the brain and around the infundibulum. This we found both in cats and rabbits. In this region of the diencephalon the ventricular wall is very thin; substances may therefore penetrate the anterior hypothalamus from the third ventricle or from the subarachnoid space.

Pickering: And does it get to the cells in the brain through the

Virchow-Robin spaces surrounding the blood vessels which are continuous with the subarachnoid?

Feldberg: If you inject a dye into the cerebral ventricles it not only stains the ventricular surface but passes through the lateral recesses into the subarachnoid space and stains the outer ventral surface of the brain, particularly around the infundibulum. And when sections are made of this region, it is found that the dye penetrates at least 1 mm.

Cooper: I have injected washed red cells into the anterior hypothalamus and pre-optic region of rabbits, and they did not have fever, but when I injected small numbers of white cells they did get fever.

Pickering: Were these untreated white cells from the same species?

Cooper: Yes, but from different animals.

Myers: The sites most sensitive to 5-HT are often close to the wall of the third ventricle as shown in Fig. 1, which illustrates six conventional stereotaxic reconstructions in coronal planes. These data are based on 32 monkeys (Myers and Yaksh, 1969). In the anterior-posterior planes, 17·0 and 18·0 mm anterior to stereotaxic zero, the sites at which 5-HT evoked a rise in temperature of 0·5°C or more are found as far as 3 mm from the ventricular lumen.

Pickering: Where is stereotaxic zero?

Myers: It is the imaginary plane between the ears. Between the posterior area and the mid-portion of the hypothalamus very few sites are sensitive to microinjections of 5-HT. In the anterior region, along the walls of the third ventricle, 5-HT produces fever. With an intracerebral haemorrhage, you can see that the blood would not have to penetrate very far to cause this hyperthermia. Just a small amount of 5-HT present in the platelets of the serum could easily cause the fever.

Cooper: One could speculate a great deal about the nature of the substance. When you let blood into a place where it does not belong you rapidly get kinin formation and all sorts of physiological garbage is formed.

Saxena: I would be glad to have your opinion on hyperthermia in heat-stroke.

Cranston: Heat-stroke appears to be failure of sweating. Bannister (1960) suggested that you could induce a syndrome not unlike heat-stroke by giving small doses of bacterial pyrogen to people in a very hot environment.

Bondy: It may be a matter of elevating the brain temperature so much that there is irreversible damage. I have seen this happen with a schizophrenic person who was given electroshock in a very hot environment; his body temperature rose to about 107°F before the fever was recognized and then it never came down. And in people exercising in severe heat

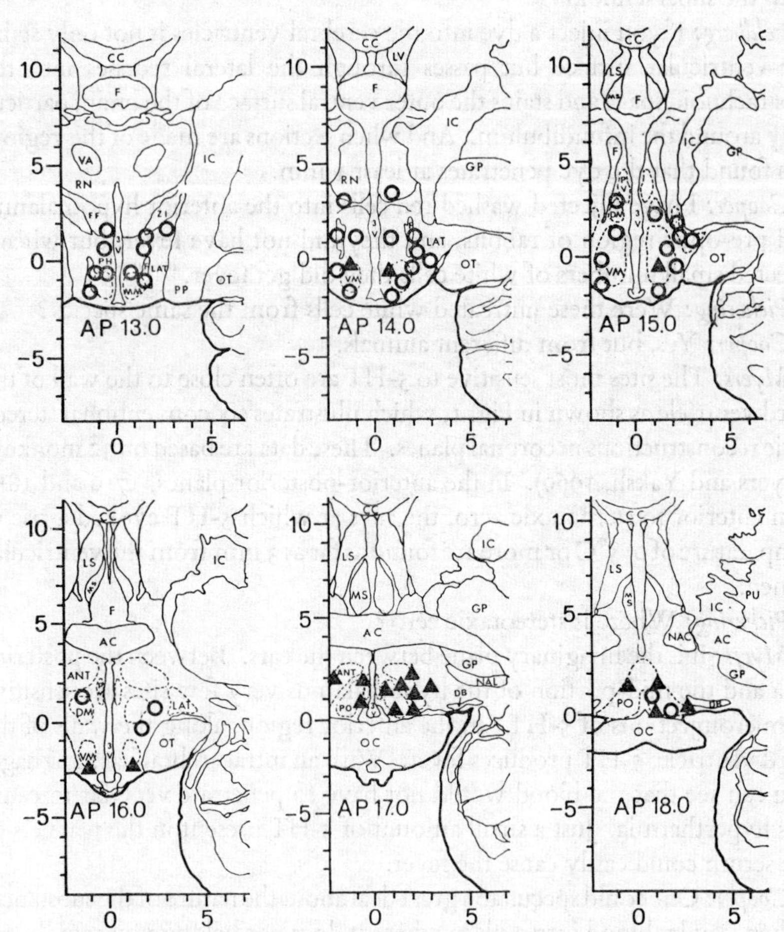

FIG. 1 (*Myers*). Anatomical "mapping" at six coronal sites (anterior-posterior plane) in the hypothalamus at which microinjections of 2–10 µg 5-HT produce hyperthermia (▲). O: sites at which injections of 5-HT caused no change in temperature; AC: anterior commissure; ANT: anterior hypothalamic area; AP: anterior-posterior plane; CC: corpus callosum; DB: diagonal band of Broca; DM: dorsomedial nucleus; F: fornix; FF: fields of Forel; GP: globus pallidus; IC: internal capsule; LAT: lateral hypothalamus; LS: lateral septal nucleus; LV: lateral ventricle; MM: mammillary body; MS: medial septal nucleus; NAC: nucleus accumbens; OC: optic chiasm; OT: optic tract; PH: posterior hypothalamic area; PO: pre-optic area; PP: cerebral peduncle; PU: putamen; PV: paraventricular nucleus; RN: reticular nucleus of the thalamus; VA: antero-ventral nucleus of the thalamus; ZI: zona incerta; 3V: third ventricle. Horizontal and lateral scales are in millimetres. Vertical zero represents the stereotaxic zero plane 10 mm above the inter-aural line. (From Myers and Yaksh, 1969.)

body temperature can rise high enough to cause permanent damage to their temperature regulatory centres (Kuno, 1956).

Myers: Gibbs (1912) exposed three monkeys to the tropical sun. He recorded subdermal temperatures as high as 48 to 50°C after only 30–40 minutes exposure. After two of the monkeys died, he did post-mortem examinations of their brains in the field and found wide-spread cerebral haemorrhages.

Whittet: This has happened to athletes who take amphetamine. They get hyperthermia and cerebral haemorrhage.

Bondy: About 10 years ago we described a couple of patients in whom fever of unknown origin was found to be associated with high levels of plasma aetiocholanolone (Bondy, Cohn and Gregor, 1965). More recently we looked at patients with unexplained fever and found no free aetiocholanolone. But the situation is further confused because some people have found high concentrations of aetiocholanolone in some patients with unexplained fever, but not necessarily in association with a fever. I agree with Dr Cranston that if this syndrome does exist, then it must be a very rare situation. There are some bile acids (e.g. lithocholic acid) which are pyrogenic and the search for these as the possible cause of fever has really not been undertaken.

Cranston: George and co-workers (1969) found that people with fever of unknown origin had slightly, but significantly, higher plasma levels of free aetiocholanolone at all times than the control population.

Bondy: It was also suggested by Mellinkoff, Schwabe and Lawrence (1960) that a certain type of familial Mediterranean fever could be exacerbated by feeding patients a high fat diet, which suggested that the secretion of bile acid might be involved.

REFERENCES

Atkins, E., and Huang, W. C. (1958). *J. exp. Med.*, **107**, 403.
Atkins, E., and Wood, W. B. (1955). *J. exp. Med.*, **101**, 509.
Bannister, R. G. (1960). *Lancet*, **2**, 118.
Bennett, I. L., and Beeson, P. B. (1953). *J. exp. Med.*, **98**, 477–493.
Berman, M. C., Harrison, G. G., DuToit. P., Bull, A. B., and Kench, J. E. (1969). *S. Afr. med. J.*, **43**, 545–456.
Bondy, P. K., Cohn, G. L., and Gregor, P. B. (1965). *Medicine, Baltimore*, **44**, 249–262.
Bornstein, D. L., Bredenberg, C., and Wood, W. B. (1963). *J. exp. Med.*, **117**, 349–364.
George, J. M., Wolff, S. M., Diller, E., and Bartter, F. C. (1969). *J. clin. Invest.*, **48**, 558–563.
Gibbs, H. D. (1912). *Philipp. J. Sci.*, **137**, 91–112.
Gillman, S. M., Bornstein, D. L., and Wood, W. B. (1961). *J. exp. Med.*, **114**, 729.
Herion, J. C., Walker, R. I., and Palmer, J. G. (1961). *J. exp. Med.*, **113**, 1115.

KUNO, Y. (1956). *Human Perspirations*. Springfield, Ill.: Thomas.
MELLINKOFF, S., SCHWABE, A. D., and LAWRENCE, J. S. (1960). *Trans. Ass. Am. Physns*, **73**, 197–205.
MYERS, R. D., and YAKSH, T. L. (1969). *J. Physiol., Lond.*, **202**, 483.
SNELL, E. S., and ATKINS, E. (1967). *Am. J. Physiol.*, **212**, 1103.
WHITTET, T. D. (1964). In *Proc. XXIII Int. pharm. Congr.*, Münster, 1963, pp. 467–474. Frankfurt: Govi-Verlag.

THE MECHANISM OF ACTION OF ANTIPYRETICS

M. D. Rawlins, C. Rosendorff, and W. I. Cranston

Department of Medicine, St. Thomas's Hospital Medical School, London

Salicylates have been used in clinical medicine for many centuries. Their use as antipyretics dates from the end of the eighteenth century (Bywaters, 1963) and they are still widely used for this purpose. However, despite their popularity amongst clinicians, the mechanism by which salicylates lower temperature during fever is poorly understood. The work of Barbour and Herriman (1920) and Guerra (1944) suggested that salicylates might exert their antipyretic effect by acting on the anterior hypothalamus to increase plasma volume, but the experiments on which these conclusions were based are by modern standards unconvincing. This and the advances in our knowledge of temperature regulation and pathogenesis of fever over the last 25 years have led to renewed interest in the mechanism of the action of antipyretic drugs. Most attention has been given to salicylates, which will be the only antipyretics considered here.

Rosendorff and Cranston (1968) showed that intravenous sodium salicylate given to patients with naturally occurring fever produced rapid and progressive defervescence which started within 5–15 minutes and was associated with an increase in cutaneous heat loss. In addition, they demonstrated that salicylates had no effect on either the temperature or the thermoregulatory reflexes of afebrile subjects. It is reasonable, therefore, to hypothesize that salicylates interfere at one or more points in the process by which fever is produced (Fig. 1).

They could theoretically act by:

(1) Interfering with the release or formation of endogenous pyrogen by leucocytes or other cells;
(2) Inactivating circulating endogenous pyrogen;
(3) Interfering with the effector pathways or end-organs responsible for producing fever (shivering, cutaneous vasoconstriction and diminished sweat production);
(4) Interfering with the action of endogenous pyrogen in the central nervous system.

These possibilities will be considered separately.

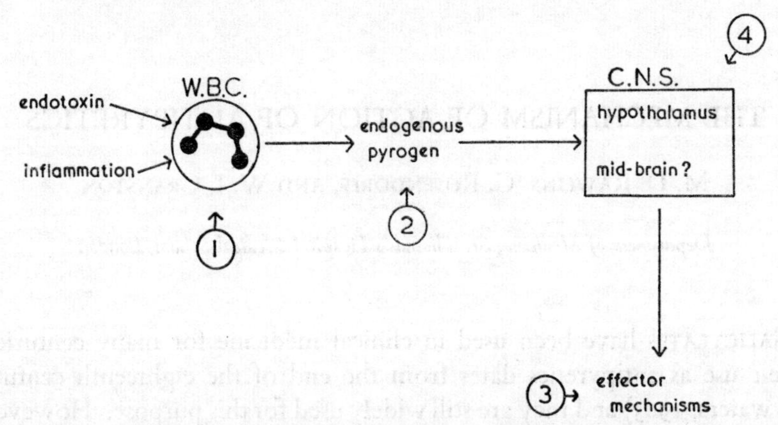

FIG. 1. Diagrammatic representation of the pathogenesis of fever. The encircled numbers indicate the points at which salicylates might theoretically act to produce antipyresis.

INTERFERENCE WITH ENDOGENOUS PYROGEN FORMATION OR RELEASE

Gander, Chaffee and Goodale (1967) showed *in vitro* that salicylates reduced the yield of endogenous pyrogen from rabbit leucocytes incubated with endotoxin, and this has subsequently been confirmed by Grundman (1969). Gander, Chaffee and Goodale (1967) also demonstrated that pretreating rabbits with salicylate reduced the febrile response to an intravenous injection of endotoxin but not to an intravenous injection of endogenous pyrogen. On this evidence they suggested that salicylates acted as antipyretics solely by interfering with the release of endogenous pyrogen from cells.

The hypothesis has been tested in normal human volunteers by inducing a steady-state fever by an intravenous priming injection followed by a sustaining infusion of autologous plasma containing endogenous pyrogen (Adler *et al.*, 1969). If salicylates act solely by stopping the production or release of endogenous pyrogen from "activated" cells then they should have no effect on this experimental fever if the infusion of endogenous pyrogen is continued while the salicylate is given; conversely, stopping the infusion of endogenous pyrogen should be followed by prompt defervescence. In fact the opposite occurred (Fig. 2): an intravenous injection of sodium salicylate (2 g), given during a steady-state fever and while the infusion of endogenous pyrogen continued, was followed by a rapid and progressive fall in temperature over the succeeding 50 minutes, but no significant changes in temperature were observed during this period if the infusion of endogenous pyrogen was stopped. These experiments indicate that salicylates do not act only by preventing the release of endogenous pyrogen from cells.

In the rabbit too it is possible to produce a steady-state fever by a priming injection followed by a sustaining infusion of homologous endogenous pyrogen, and this fever appears to persist for as long as the infusion is continued (Cranston *et al.*, 1970*b*). Sodium salicylate given intravenously

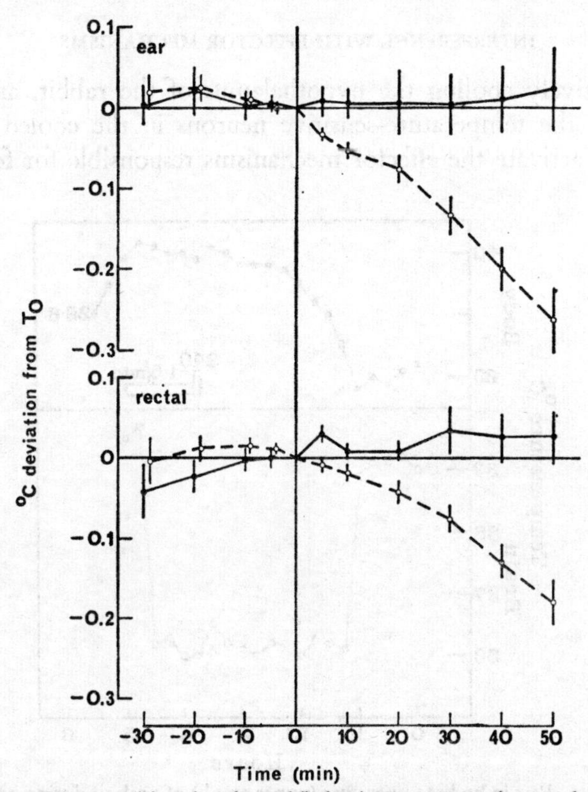

FIG. 2. Mean ear (above) and rectal (below) temperature changes during pyrogen-induced fever in a series of normal volunteers. At time zero (T_0) either 2 g sodium salicylate were given intravenously, followed by a sustaining infusion of 20 mg/min (open circles), or the pyrogen infusion was stopped (closed circles). Ordinate: the difference between the temperature at T_0 (°C) and the preceding and subsequent temperatures. Vertical lines represent ± 1 S.E.M. (From Adler *et al.* 1969, reproduced by kind permission of the Editor of *Clinical Science.*)

four hours after the start of the infusion, and while the infusion continued, produced rapid and progressive defervescence which was dose-dependent. As in man, intravenous salicylate was without effect on the temperature of afebrile rabbits.

INACTIVATION OF CIRCULATING ENDOGENOUS PYROGEN

It is conceivable that salicylates produce antipyresis by inactivating circulating endogenous pyrogen. Grundman (1969) has shown that the pyrogenicity of endogenous pyrogen is unaffected by incubation *in vitro* with salicylate. However, his experiments do not exclude the possibility that salicylates might reversibly inactivate circulating endogenous pyrogen.

INTERFERENCE WITH EFFECTOR MECHANISMS

By selectively cooling the hypothalamus of the rabbit, and thereby stimulating the temperature-sensitive neurons in the cooled area, it is possible to activate the effector mechanisms responsible for fever in the

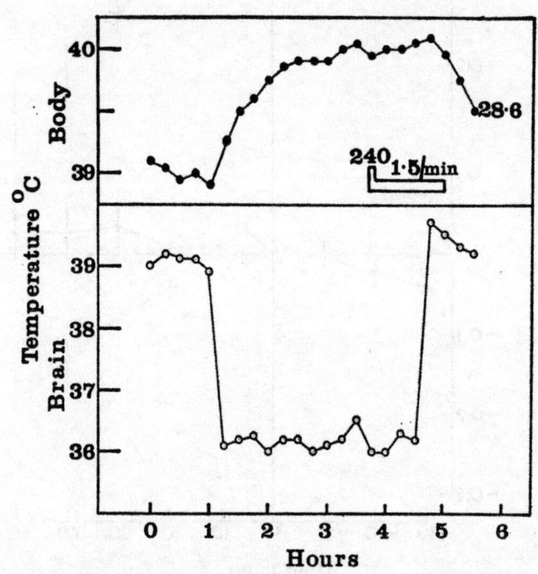

FIG. 3. Rise in body temperature (upper graph) of a rabbit during brain cooling (lower graph). The salicylate injection (240 mg) and the duration of the infusion (1·5 mg/min) are indicated by the bar in the upper graph. (From Cranston *et al.* 1970*a*, reproduced by permission of the Editor of the *Journal of Physiology*.)

absence of endogenous pyrogen. The raised body temperature which this manoeuvre produces persists for as long as the hypothalamus is cooled (Cranston *et al.*, 1970*a*). Intravenous sodium salicylate given during this period (at a dose that lowers temperature during pyrogen-induced fever) has no effect on temperature (Fig. 3).

From evidence obtained in rabbits (Cooper, Cranston and Honour, 1967) and cats (Jackson, 1967) it is likely that endogenous pyrogen acts on central thermoregulatory mechanisms within the hypothalamus and perhaps the mid-brain (Rosendorff, Mooney and Long, 1970). The temperature rises following brain cooling and administration of endogenous pyrogen are probably mediated by the same peripheral effector mechanisms. The details of the central efferent pathways responsible for both varieties of fever are poorly understood, but it is reasonable to suppose that they share a common anatomical pathway within the central nervous system below the mid-brain. The absence of any antipyretic response to salicylate during the fever produced by hypothalamic cooling makes it unlikely that salicylates act peripherally or through the common pathway within the central nervous system.

By a process of exclusion, therefore, it seems likely that salicylates exert at least part of their antipyretic effect by acting on the central nervous system.

DIRECT ACTION ON THE CENTRAL NERVOUS SYSTEM

(a) *Site of action*

Sodium salicylate injected into a lateral cerebral ventricle during a steady-state fever, and four hours after a priming injection followed by a sustaining infusion of homologous endogenous pyrogen, produces, in the rabbit, a dose-dependent fall in temperature (Fig. 4) (Cranston *et al.*, 1970*b*). The doses of salicylate given intraventricularly were less than one per cent of the systemic dose that produces antipyresis. However, whereas febrile animals receiving salicylate intravenously show a sustained fall in temperature during the subsequent 80 minutes, febrile animals receiving salicylate intraventricularly show a fall in temperature for up to only 40 minutes, followed by a gradual rise towards the control temperature. We have attributed this phenomenon to clearance of salicylate from the nervous system. The same intraventricular doses of salicylate had no effect on the temperature of afebrile rabbits.

These experiments established that part of the antipyretic property of salicylates is mediated via the central nervous system. To localize this further, rabbits with a steady-state fever induced by an intravenous infusion of endogenous pyrogen were bilaterally microinjected in various parts of the brain with 6–30 μg sodium salicylate in 10 μl of mock cerebrospinal fluid. Only one site was studied in any one animal, and at the end of each experiment 10 μl of indian ink were injected so that the site could be accurately located. The results of these experiments are summarized in Fig. 5: the two areas where antipyretic responses to microinjections of salicylate could be consistently demonstrated were the pre-optic hypothalamus and the mid-brain. Both these areas of the central nervous system appear to contain thermosensitive neurons (Hellon, 1967; Nakayama and Hardy, 1969), and to respond to local injections of endogenous

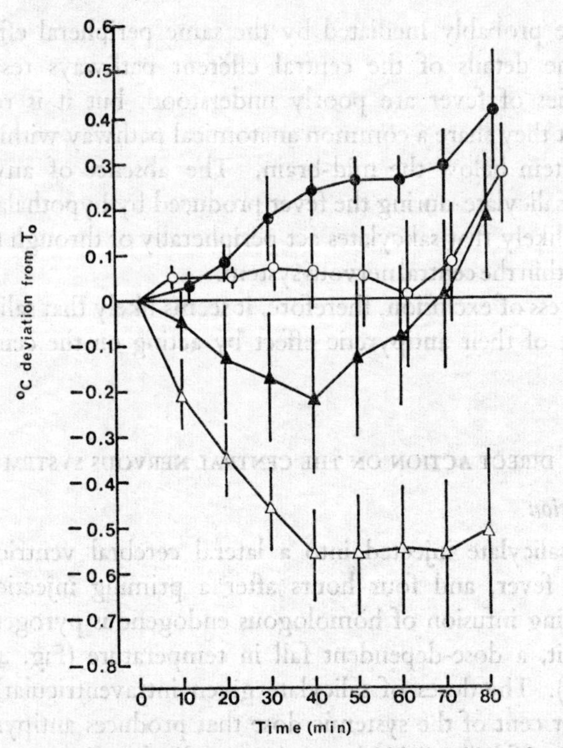

FIG. 4. Mean changes in rectal temperature of febrile rabbits after intraventricular salicylate injection. All animals had received a priming injection followed by a sustaining infusion of endogenous pyrogen for four hours before the salicylate was given, and the infusion of endogenous pyrogen was continued throughout the experiment. Ordinate: the difference between the temperature at T_0 (when the salicylate was given) and subsequent temperatures. Vertical lines represent ± 1 S.E.M.

●——●: 0·4 ml mock cerebrospinal fluid control; ○——○: 0·12 mg sodium salicylate; ▲——▲: 0·60 mg sodium salicylate; △——△: 1·20 mg sodium salicylate.

pyrogen by producing a rise in temperature. Furthermore, the pre-optic hypothalamus contains high concentrations of monoamines which are thought to be involved in thermoregulation (Feldberg, 1968), and the mid-brain contains cell bodies whose axons reach the pre-optic hypothalamus and are rich in 5-hydroxytryptamine (Dahlstrom and Fuxe, 1964).

(b) *Mechanism of action*

Cooper, Grundman and Honour (1968) suggested that salicylates might act by interfering with the passage of endogenous pyrogen into the hypothalamus and pre-optic areas. They based this hypothesis on the fact that pretreatment with salicylate depresses the febrile response to intravenous endogenous pyrogen, but not to endogenous pyrogen injected

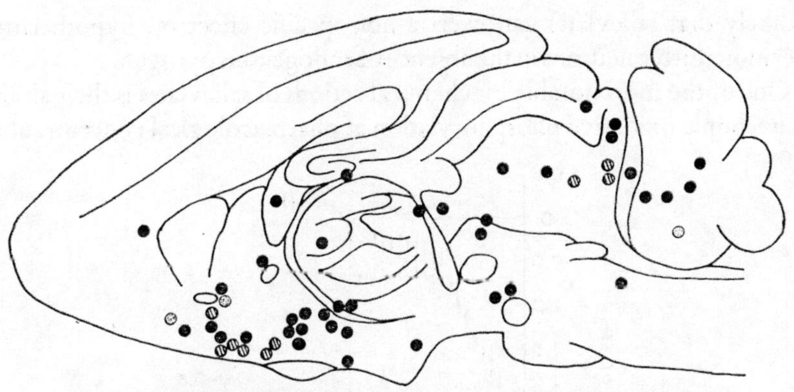

FIG. 5. Results of bilateral microinjections (10 μl) of 6–30 μg sodium salicylate in rabbits with fever induced by an infusion of endogenous pyrogen. Results projected onto a sagittal section of the rabbit brain. The two areas which give antipyretic responses are in the pre-optic hypothalamus and the mid-brain. ●: no response; ◍: doubtful response; ◍: response to 30 μg sodium salicylate or less.

into the cerebral ventricles of rabbits. We have tested this hypothesis by studying the temperature responses of rabbits to intraventricular injections of endogenous pyrogen. Animals injected intraventricularly with 100 μl of autologous plasma containing endogenous pyrogen show a rise in rectal temperature which reaches a plateau 180–240 minutes after the injection (Fig. 6*a*). It was found that both intraventricular (Fig. 6*b*) and intravenous salicylate (Fig. 6*c*) produced defervescence when given 180 minutes after the injection of endogenous pyrogen (Cranston *et al.*, 1970*a*). It is assumed that in these experiments intraventricularly injected endogenous pyrogen gained access to the hypothalamus without crossing the blood-brain barrier. The fact that the fever produced by this manoeuvre is affected by both systemic and local salicylate is incompatible with the hypothesis of Cooper, Grundman and Honour (1968).

Wit and Wang (1968) and Eisenman (1969) have studied the firing rates of thermosensitive neurons in the pre-optic and anterior hypothalamus before and after endotoxin was intravenously injected into anaesthetized cats. They both found a decrease in the sensitivity of warmth-sensitive neurons after the administration of endotoxin, and Eisenman also observed a rise in sensitivity of cold-sensitive neurons without any change in their firing rate at normal core temperature. Wit and Wang observed that the reduced sensitivity of warmth-sensitive neurons after the injection of endotoxin was restored by both intracarotid and intravenous sodium acetyl salicylate. How this effect is mediated, and its relevance to the mechanism of salicylate-induced antipyresis, is uncertain. However, our experiments with brain-cooling described above (Cranston *et al.*, 1970*a*) make it unlikely that salicylates can exert a non-specific effect on hypothalamic thermosensitive neurons in the absence of endogenous pyrogen.

One of the most notable biochemical actions of salicylates is their ability to uncouple oxidative phosphorylation at pharmacological concentrations

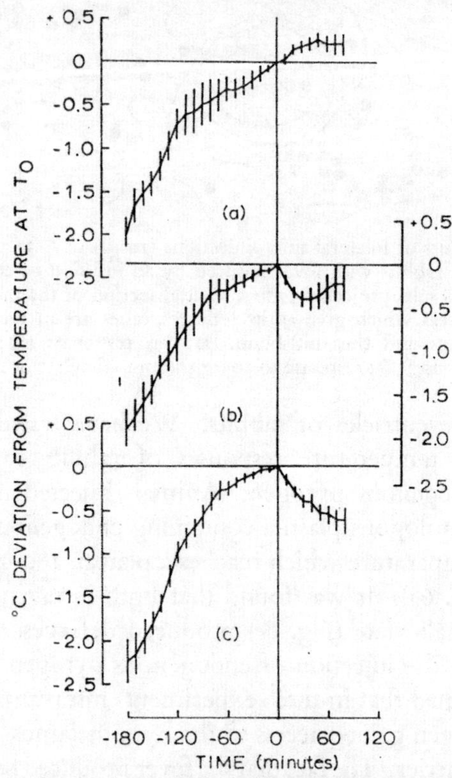

FIG. 6. Mean changes in rectal temperature during fever induced by an intraventricular injection of 100 μl autologous plasma containing endogenous pyrogen given 3 hours before T_0. Ordinate: the difference between the temperature at T_0 and all preceding and subsequent temperatures. (a) Results from six rabbits receiving endogenous pyrogen only; (b) results from five rabbits receiving 1·2 mg sodium salicylate intraventricularly at T_0; (c) results from six rabbits receiving 360 mg sodium salicylate intraventricularly at T_0 followed by an infusion of 2·0 mg/min. Vertical lines represent ±1 S.E.M. (From Cranston *et al.*, 1970*a*, reproduced by kind permission of the Editor of the *Journal of Physiology*.)

(Brody, 1956). To establish whether this property is related to the anti-pyretic effects of salicylates, febrile rabbits were given 2,4-dinitrophenol by intraventricular injection four hours after the start of an endogenous pyrogen infusion. No fall in temperature was observed after the ventricular injection.

One intriguing problem that has arisen during recent studies on the mechanism of salicylate-induced antipyresis has been the discrepancies between the results of workers who have administered salicylates before the production of fever and those who have given salicylates during an established fever.

Gander, Chaffee and Goodale (1967) were unable to reduce the febrile response to an intravenous injection of endogenous pyrogens by pretreating

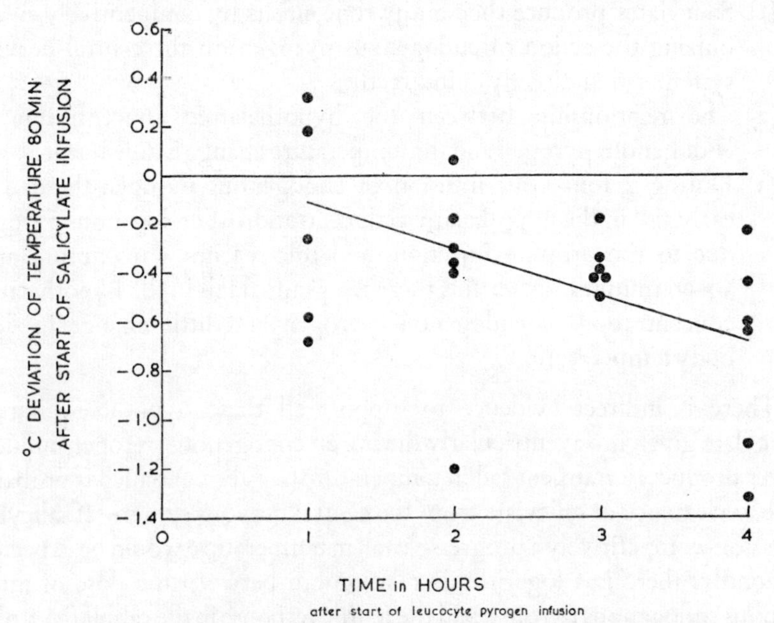

FIG. 7. Changes in rectal temperature (ordinate) 80 minutes after an intravenous injection of 240 mg sodium salicylate and an infusion of 1·5 mg/min in 4 groups of rabbits who received their salicylate at either 1, 2, 3 or 4 hours after the start of an infusion of endogenous pyrogen. The regression line is also shown.

rabbits with salicylate, and Grundman (1969) observed a similar phenomenon when he used a small series of rabbits; he had to use a much larger series in a "cross-over" experiment to demonstrate an antipyretic effect. Rosendorff and Cranston (unpublished observations) found in a few experiments on human subjects that pretreatment with salicylate had no effect on the temperature responses to autologous plasma containing endogenous pyrogen. On the other hand, salicylate given during an established fever, induced by a priming injection followed by a sustaining infusion of endogenous pyrogen, produced rapid and progressive defervescence in both man and the rabbit. To study this problem further we have observed the effect of salicylate given intravenously to rabbits at one, two, three and four hours after a priming injection and infusion of endogenous pyrogen. It is clear from Fig. 7 that salicylates have little antipyretic effect when given one hour after the start of an infusion of endogenous pyrogen, but that the magnitude of the antipyretic response increases with the duration of the infusion. This is compatible with the finding that salicylate given before a single injection of endogenous pyrogen has little effect on the subsequent febrile response. To explain these findings we have suggested that:

(1) Salicylates produce their antipyretic effects by competitively antagonizing the action of endogenous pyrogen on the central nervous system, either directly or indirectly;
(2) The relationship between the hypothalamic concentration of endogenous pyrogen and the temperature change is non-linear;
(3) During a four-hour infusion of endogenous pyrogen there is an early rise in the hypothalamic concentration of endogenous pyrogen due to the priming injection, and this reaches a maximum after 30–90 minutes. After this there is a gradual fall in the hypothalamic concentration of endogenous pyrogen and little change in deep body temperature.

There is indirect evidence to support all these hypotheses. Firstly salicylate given intraventricularly during an endogenous pyrogen-induced fever produces a transient fall in temperature (see above) which is probably due to clearance of salicylate from the central nervous system. If salicylate acts non-competitively a progressive fall in temperature would be expected. Secondly, there is a logarithmic relationship between the dose of intravenous endogenous pyrogen and the febrile response in the rabbit (Murphy, 1966), and it is reasonable to suppose that a similar relationship might hold between the hypothalamic concentration of endogenous pyrogen and the febrile response. In the absence of sufficiently sensitive techniques for measuring hypothalamic endogenous pyrogen concentration during steady-state fever induced by a priming injection and sustaining infusion of endogenous pyrogen, we have measured the concentration of transferable pyrogen in the plasma of animals bled at one, two and four hours after the start of an infusion of endogenous pyrogen. We used four groups of rabbits; each group received either a control infusion of normal rabbit plasma for two and a half hours, or an endogenous pyrogen infusion for one, two or four hours. The plasma from animals in each group was pooled and assayed in a further group of eight rabbits in a cross-over study. The results (Fig. 8) show a gradual fall in transferable pyrogen during the course of a four-hour infusion of endogenous pyrogen which represents a fall in circulating endogenous pyrogen of 50 per cent, and it is possible that a similar fall might occur in the concentration of hypothalamic endogenous pyrogen. These results indicate that the increasing magnitude of the antipyretic response to intravenous salicylate during an endogenous pyrogen infusion might be due to this drug changing the effective concentration of endogenous pyrogen in the hypothalamus. Due to the non-linearity of

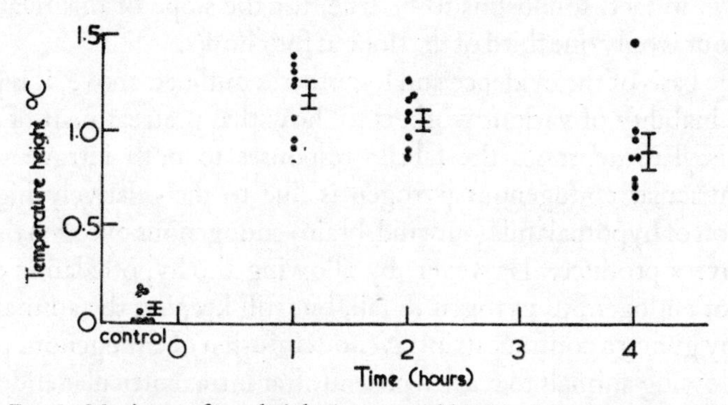

FIG. 8. Maximum fever height in assay-rabbits receiving 20 ml plasma from rabbits given a priming injection followed by a sustaining infusion of endogenous pyrogen for 1, 2, and 4 hours. The control injection was of 20 ml plasma from rabbits receiving a continuous infusion of normal rabbit plasma for 2·5 hours.

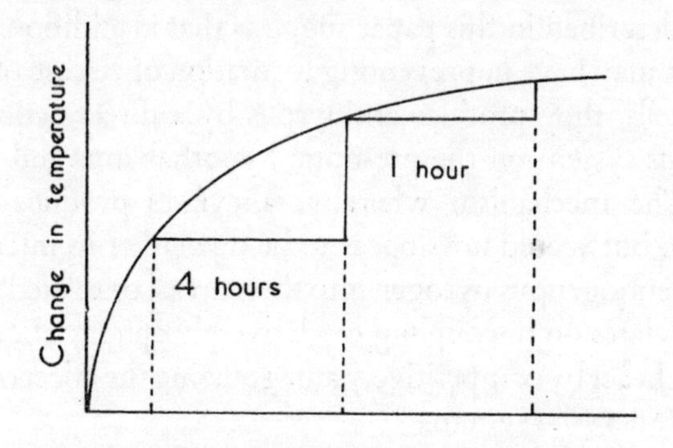

FIG. 9. Hypothetical relationship between the hypothalamic concentration of endogenous pyrogen (EP) and the change in body temperature it induces. It is suggested that salicylates might act by reducing the effective hypothalamic concentration of endogenous pyrogen by competitive antagonism. With a high concentration of endogenous pyrogen in the hypothalamus (1 hour after the start of an intravenous infusion of endogenous pyrogen) salicylates would have little effect on temperature. With a lower concentration of endogenous pyrogen in the hypothalamus (4 hours after the start of an intravenous infusion of endogenous pyrogen) the effect would be considerably greater.

the concentration-response relationship, salicylate would be less effective at one hour (when a higher concentration of endogenous pyrogen is present in the hypothalamus) than at four hours (Fig. 9). If this is correct, then due to the increased concentration of hypothalamic endogenous pyrogen at one hour, the slope of the dose-response relationship between intraventricular salicylate and the fall in temperature should be less at one hour than at four hours after the start of an endogenous pyrogen infusion. We have, in fact, found this to be true, for the slope of this relationship at one hour is only one third of the slope at four hours.

On the basis of the evidence and hypotheses outlined above, it is possible that the inability of various workers to show that pretreatment of animals with salicylate depresses the febrile responses to both intravenous and intraventricular endogenous pyrogen is due to the relatively high concentration of hypothalamic (and mid-brain) endogenous pyrogen that these manoeuvres produce. However, by allowing the hypothalamic concentration of endogenous pyrogen to fall, but still keeping the animal febrile (either by giving a continuous intravenous infusion of endogenous pyrogen or by allowing animals to reach a plateau after intraventricular endogenous pyrogen), salicylate-induced antipyresis can be demonstrated.

CONCLUSIONS

The work described in this paper suggests that in addition to any action that salicylates may have in preventing formation or release of endogenous pyrogen by cells, they produce antipyresis by a direct action within the central nervous system on the pre-optic hypothalamus and probably the mid-brain. The mechanism whereby salicylates produce this effect is still unknown, but would not appear to be due either to interference with the passage of endogenous pyrogen into these areas, or related to the known effects of salicylates on uncoupling oxidative phosphorylation. However, salicylates might act by competitively antagonizing the effect of endogenous pyrogen on the nervous system.

SUMMARY

The mechanism of action of antipyretics in general is poorly understood. Salicylates are one of the drugs used most widely in clinical medicine to induce defervescence in febrile patients, but they have no effect upon either the temperature or thermoregulatory reflexes of afebrile subjects. It is therefore reasonable to suggest that salicylates might act at one or more steps in the process by which fever is produced.

While there is evidence that salicylates reduce the yield of endogenous pyrogen from cells, this is not the sole mechanism by which they exert their antipyretic effect. In both man and rabbits with steady-state fevers induced by a continuous intravenous infusion of endogenous pyrogen, intravenous salicylate produces defervescence while the endogenous pyrogen infusion is continued.

There is no evidence that salicylates either inactivate circulating endogenous pyrogen or interfere with the effector mechanisms responsible for fever.

In rabbits with a steady-state fever induced by an intravenous infusion of endogenous pyrogen, small amounts of salicylate injected into a lateral ventricle produce dose-dependent defervescence. This effect appears to be mediated by a direct action of salicylate on both the hypothalamus and the mid-brain, and it is not due to interference by salicylate with the passage of endogenous pyrogen into these areas of the brain, or to uncoupling of oxidative phosphorylation.

It is suggested that salicylates antagonize the action of endogenous pyrogen within the central nervous system competitively.

REFERENCES

ADLER, R. D., RAWLINS, M., ROSENDORFF, C., and CRANSTON, W. I. (1969). *Clin. Sci.*, **37**, 91–97.

BARBOUR, H. G., and HERRIMAN, J. B. (1920). *Proc. natn. Acad. Sci. U.S.A.*, **6**, 136–139.

BRODY, T. M. (1956). *J. Pharmac., exp. Ther.* **117**, 39–51.

BYWATERS, E. G. L. (1963). In *Salicylates* (An International Symposium), pp. 8–12, ed. Dixon, A. St. J., Martin, B. K., Smith, M. J. H., and Wood, P. H. N. Boston: Little, Brown.

COOPER, K. E., CRANSTON, W. I., and HONOUR, A. J. (1967). *J. Physiol., Lond.*, **191**, 325–337.

COOPER, K. E., GRUNDMAN, M. J., and HONOUR, A. J. (1968). *J. Physiol., Lond.*, **196**, 56–57P.

CRANSTON, W. I., HELLON, R. F., LUFF, R. H., RAWLINS, M. D., and ROSENDORFF, C. (1970*a*). *J. Physiol., Lond.*, **210**, 593–600.

CRANSTON, W. I., LUFF, R. H., RAWLINS, M. D., and ROSENDORFF, C. (1970*b*). *J. Physiol., Lond.*, **208**, 251–259.

DAHLSTROM, A., and FUXE, K. (1964). *Acta physiol. scand.*, **62**, suppl. 232.

EISENMAN, J. S. (1969). *Am. J. Physiol.*, **216**, 330–335.

FELDBERG, W. (1968). In *Recent Advances in Pharmacology*, pp. 349–397, ed. Robson, J. M., and Stacey, R. S. London: Churchill.

GANDER, G. W., CHAFFEE, J., and GOODALE, F. (1967). *Proc. Soc. exp. Biol. Med.*, **126**, 205–209.

GRUNDMAN, M. J. (1969). D. Phil. thesis, University of Oxford.

GUERRA, F. (1944). *J. Pharmac. exp. Ther.*, **82**, 103–109.

HELLON, R. F. (1967). *J. Physiol., Lond.*, **193**, 381–395.

JACKSON, D. L. (1967). *J. Neurophysiol.*, **30**, 586–602.

MURPHY, P. A. (1966). D.Phil. thesis, University of Oxford.

NAKAYAMA, T., and HARDY, J. D. (1969). *J. appl. physiol.*, **27**, 848–857.

ROSENDORFF, C., and CRANSTON, W. I. (1968). *Clin. Sci.*, **35**, 81–91.

ROSENDORFF, C., MOONEY, J. J., and LONG, C. N. H. (1970). *Fedn Proc. Fedn Am. Socs exp. Biol.*, **29**, 523, abst. 1547.

WIT, A., and WANG, S. C. (1968). *Am. J. Physiol.*, **215**, 1160–1169.

DISCUSSION

Grundman: When I pretreated rabbits with an intravenous injection of 300 mg sodium salicylate and half an hour later gave a dose of endogenous pyrogen there was a considerable antipyretic effect—a reduced febrile response of about 50 per cent (Cooper, Grundman and Honour, 1968; Grundman 1969, pp. 74–75). This was hard to show in individual rabbits because the same rabbit may respond differently to one dose of pyrogen on different days and different rabbits may respond differently to one dose on the same day. If we just compared single rabbits on different days we quite often got a higher febrile response in the salicylate-treated animal.

Rawlins: I am sure this is true. However, in a few experiments in man we found it impossible to demonstrate any antipyretic effect when subjects were pretreated with salicylate and then given a single intravenous injection of leucocyte pyrogen.

Cranston: As competitive antagonism this would make perfectly good sense.

Grundman: We followed this experiment by pretreatment with salicylate followed by an intraventricular dose of leucocyte pyrogen (Grundman, 1969, pp. 81–84). Dr Rawlins, you postulated that our concentration of leucocyte pyrogen must have been higher on this occasion than it had been with the intravenous dose, and that this was one reason why we showed no difference between the group which had been pretreated with salicylate and the group which had not. But our mean fever with intraventricular leucocyte pyrogen varied between 0·4 and 0·6°C, which is substantially less than yours, and in fact less than the fevers we were getting with intravenous leucocyte pyrogen (0·7 to 1·0°C).

Cooper: Dr Rawlins, your animals had a fever of about 2·5°C. By giving such huge doses of pyrogen intraventricularly you might be getting an inflammatory response; in this situation the salicylate might get in and out more rapidly. Despite this, your hypothesis seems more likely to be correct than the one we originally put forward (Cooper, Grundman and Honour, 1968), which itself may be part of the answer.

Cranston: This is a very reasonable criticism and there is still a lot wrong with this hypothesis. Dr Rawlins is suggesting that the dose-effect curve flattens out at high hypothalamic pyrogen concentrations. This fits reasonably well with the dose-effect relationship between intravenous leucocyte pyrogen and the early fever peak. However at high doses, leucocyte pyrogen produces a biphasic response with a much higher second peak (Bornstein, Bredenberg and Wood, 1963).

Whittet: Dr Palmer and I gave rabbits a salicylate and other antipyretics by stomach tube and then injected quite big doses of bacterial pyrogens. We got virtually complete elimination of fever (Baker *et al.*, 1963).

Cranston: But at least two mechanisms may be working in these bacterial pyrogen experiments; you have got suppression of leucocyte pyrogen production and also a central effect of salicylates.

Myers: Have you ever microinjected the bacterial pyrogen directly into the anterior hypothalamus and then given salicylates systemically?

Rawlins: No.

Myers: When *E. coli* is microinjected into the anterior hypothalamus to ensure that there is no circulating or systemic leucocytic pyrogen, intragastric salicylate produces a dose-dependent, transient hypothermia. This would indicate that the salicylate must be acting on elements of the nervous system, specifically the hypothalamus (Myers, Rudy and Yaksh, 1971). We have also obtained results which show that in the monkey at least the salicylate microinjected into the hypothalamus produced no temperature response, or gave a slight rise in temperature. These are, of course, only preliminary results and the localization of the cannulae placements has not been completed.

Rawlins: We obtained transient falls in temperature when we microinjected small amounts of salicylates into both the pre-optic hypothalamus and the mid-brain of febrile rabbits.

Myers: I am quite sure that you have not localized the action of a salicylate by injecting a volume as devastating as 10 μl on each side of the rabbit's brain. The unilateral space occupied by the rabbit's hypothalamus is something less than 9 μl. Moreover, it has been known for at least 10 years that any volume of fluid over 4 μl injected into brain-stem tissue flows directly up the shaft and into the cerebral ventricles (see Myers and Sharpe, 1968).

Rawlins: While I accept that we do not know the volume of distribution of our microinjections, we do know that 120 μg salicylate given intraventricularly produces little change in temperature when given to animals with a steady-state fever. On the other hand 12–30 μg salicylate injected into the pre-optic region of the hypothalamus or the mid-brain produces a much greater fall in temperature.

Feldberg: You use cooling as control. I would have preferred another control because cooling affects not only the synapse but all the nerve fibres in that region. How does the salicylate affect the fever produced by noradrenaline?

Rawlins: We don't know.

Cranston: It is difficult to get reproducible results with noradrenaline. We think we might be able to do this by using the tricyclic antidepressants which block noradrenaline re-uptake across the presynaptic membrane. In the rabbit they produce a reproducible temperature rise and in the cat a fall.

Feldberg: How closely related would the effect be to that of the monoamines?

Cranston: I think imipramine is operating through locally released endogenous noradrenaline.

Feldberg: Do you get constant responses with imipramine?

Cranston: Yes.

Feldberg: This would be a nice physiological control.

Cooper: We are dealing here with salicylate. I believe that Dr Grundman and Dr Murphy both have evidence that amidopyrine lowers the body temperature in the afebrile, apparently healthy animal, so other antipyretics may have very different mechanisms of operating.

Grundman: This is true. When I injected 1 mg of sodium salicylate into the lateral ventricles of afebrile animals I obtained an immediate fall in temperature of about 0·4°C which usually lasted less than one hour. The effect was the same in animals made febrile (1·2°C) by intravenous infusion of leucocyte pyrogen (Grundman, 1969, pp. 85–88).

Bondy: Dr Rawlins, you suggested that salicylate worked by competitive inhibition; you really can't say that in a enzymological sense, but perhaps you alluded to the possibility that salicylate might be involved in some way by reversibly inactivating leucocyte pyrogen. If that is the case, this would perhaps also explain why leucocyte pyrogen release appears to be impaired *in vitro*; what may be happening is that there is less released, but what is released is distorted in some way. Could your results be explained simply on the basis of the salicylate interacting with the endogenous pyrogen temporarily or permanently within the experimental system?

Grundman: We found that leucocyte pyrogen was certainly not permanently inactivated by salicylate but I can't say what happens temporarily (Grundman, 1969, pp. 76–80).

Bondy: But even if it were removed temporarily and then drifted back into the system at a sufficiently slow rate to be removed before it reached a concentration high enough to maintain the fever, it might have the same physiological effect.

Grundman: Certainly!

Bodel: In experiments with rabbit blood leucocytes stimulated by endotoxin *in vitro*, Goodale has shown that pyrogen release appears to be suppressed with salicylate (Gander, Chaffee and Goodale, 1967). On the other hand, a medical student in our laboratory, C. Reynolds, working with human blood leucocytes stimulated by phagocytosis *in vitro*, has found very little, if any, effect of salicylate (unpublished).

Pickering: Dr Rawlins, I don't like your assumptions about the pyrogen concentration in the brain because it seems to me that the temperature would reflect it.

Cranston: Yes, but not necessarily linearly. We had to produce this model to try to explain the previous conflicting evidence, and this appears the least unlikely hypothesis to fit the facts.

Feldberg: What is the relation of these mid-brain areas to the raphe?

Rawlins: The median raphe is lower down.

Feldberg: Was there any effect on the raphe?

Rawlins: No.

Grundman: Murphy has done similar experiments to Goodale's, and has shown in rabbits that salicylate to some extent inhibits the release of leucocyte pyrogen from blood exposed to bacterial pyrogen, but this only accounted for about a quarter of the antipyretic effect (Murphy, P. A., unpublished).

Cranston: If one is considering the possibility of a reversible situation, then whether or not you dialyse the salicylate out of the rabbit or human blood at the end of the experiment might be quite important.

Bodel: I don't believe Goodale dialysed his material, so he was injecting salicylate plus pyrogen.

Rawlins: However, Gander, Chaffee and Goodale (1967) were using pharmacological concentrations of salicylate in their *in vitro* incubation and this would only represent a minute dose of salicylate when given to whole animals.

Cranston: In most of the *in vitro* experiments where antipyretics have been incubated with cells and bacterial pyrogen, the quantity of bacterial pyrogen used has been such that the leucocytes give an almost 100 per cent yield of leucocyte pyrogen. In clinical situations the position may be quite different, and larger numbers of cells of whatever kind, might be activated to a much smaller extent. The effects of antipyretics might be quite different in these circumstances.

REFERENCES

BAKER, J. A., HAYDEN, J., MARSHALL, P. G., PALMER, C. H. R., and WHITTET, T. D. (1963). *J. Pharm. Pharmac.*, **15**, 97–100T.

BORNSTEIN, D. L., BREDENBERG, C., and WOOD, W. B. (1963). *J. exp. Med.*, **117**, 349–364.

COOPER, K. E., GRUNDMAN, M. J., and HONOUR, A. J. (1968). *J. Physiol., Lond.*, **196**, 56–57P.

GANDER, G. W., CHAFFEE, J., and GOODALE, F. (1967). *Proc. Soc. exp. Biol. Med.*, **126**, 205–209.

GRUNDMAN, M. J. (1969). D.Phil. Thesis, University of Oxford.

MYERS, R. D., RUDY, T. A., and YAKSH, T. L. (1971). *Experientia*, in press.

MYERS, R. D., and SHARPE, L. G. (1968). *Physiol. Behav.*, **3**, 987.

PHARMACEUTICAL ASPECTS OF PYROGENS IN HOSPITAL AND INDUSTRY

C. H. R. PALMER

School of Pharmacy, City of Leicester Polytechnic, Leicester

SINCE the introduction of the hypodermic syringe in 1855, medicaments have been administered parenterally to patients. As early as 1863 the ill-effects which followed treatment by injection were being investigated. Billroth (1865) noted that fever appeared in patients while they were receiving a course of injections. Since then various workers have discovered the cause of the fevers and the properties of the pyrexial agents, such as heat stability (Centanni, 1894), a molecule small enough to pass through diatomite bacteria-proof filters (Hort and Penfold, 1911, 1912), and a bacterial origin (Jona, 1916). Such substances were termed "pyrogens" by Burdon-Sanderson (1876). Seibert (1923) found that these substances were produced by waterborne bacteria, and that correct distillation would eliminate pyrogens. More recently, Co Tui and Schrift (1942) suggested that airborne bacteria, as well as some yeasts and moulds, also produced pyrogens, and Bennett, Wagner and LeQuire (1949*a*, *b*) and Wagner and Bennett (1950) found that viruses too caused pyrogenic effects.

Pyrogens are microbial in origin, and their potency and properties depend on the microorganism producing these substances. The most potent pyrogen-producing organisms are the gram-negative bacteria (Wylie and Todd, 1948, 1949); gram-positive bacterial species, yeasts, moulds and viruses are less pyrogenic. Thus microbial contamination of solute(s), solvents, equipment or apparatus during processing of preparations can cause the final product to be pyrogenic.

Although some purified pyrogens have been shown to be thermolabile (Palmer and Whittet, 1961; Palmer, 1967), most others are thought to be thermostable. Because of this thermostability, sterilization of the final product by heating methods, as officially recommended by pharmacopoeias, will not destroy all the pyrogenic contaminants. Consequently, pyrogens present problems to the pharmaceutical profession in both hospitals and industry, and it must always be assumed that the pyrogenic contaminants which may be present are thermostable.

In both hospitals and industry, the number of injectable products which are prepared is increasing. Pyrogens only produce fever when they are injected into the body. Other sterile products, such as eye drops, eye ointments and eye lotions are not administered into the circulatory system, and therefore do not produce a pyrogenic effect. Subcutaneous or intramuscular injections of small volume, even if they are strongly pyrogenic, do not cause a pronounced febrile response in the patient. With intramuscular injections, some rise in body temperature may ensue, but this might be due to the irritant nature of the substance itself, such as sulphur in oil, or to the pyrexial nature of the chemical substances, for example certain steroids, native dextrans, tetrahydro-β-naphthylamine, 2,4-dinitrophenol, lysergide, methyl cellulose, kaolin, 5-hydroxytryptophan and pentachlorophenol. These materials produce a febrile response similar to a pyrogenic response, but this is due to the temperature-raising property of the substances and not to microbial contamination of the product. Intravenous injections of small volume must have a very high pyrogenic content to produce a febrile response in the patient. Infusion fluids are the most likely substances to cause a pyrogenic response because of the large volume injected.

The pharmacopoeial tests for pyrogens are "limit" tests performed on rabbits and are described in the *British Pharmacopoeia* (1968), the *United States Pharmacopoeia* (1955) and the *International Pharmacopoeia* (1955). The samples to be tested are large compared with the amount normally injected into a patient. Further, the rabbit is more sensitive to pyrogens than are humans, and consequently these pharmacopoeial tests are satisfactory for the detection of pyrogens.

Because so many injections are now prepared in hospital departments, it is impracticable to test every batch for the presence of pyrogens. Higher priority should therefore be given to testing those products where pyrogenic contamination is most likely to occur. Charlton (1965), referring to the pyrogen testing of radiopharmaceuticals, suggested that products be classified in two general classes:

(i) *High risk*, which includes:
- (a) products administered in large volumes e.g. infusion fluids.
- (b) products which are good media for microbial growth, or contain substances which may themselves be pyrogenic.

(ii) *Low risk*, which includes:
- (a) products administered in a small volume.
- (b) products which either contain substances inhibitory to bacterial growth, or are themselves inhibitory, so forming a poor microbial growth-medium.

Products in these two classes differ greatly in the degrees of danger presented. The pharmacopoeias generally take account of this difference when specifying that a pyrogen test be carried out on the product. The 1968 edition of the *British Pharmacopoeia* contains 152 monographs for various injections, 33 of which require pyrogen tests. Of the latter, several are large-volume infusion fluids, such as sodium chloride or dextrose injections, and some, such as calcium gluconate, heparin and injections containing some antibiotics, are possibly particularly prone to pyrogen contamination.

There appear to be three ways of handling the large number of injections prepared in hospitals for which routine pyrogen testing is impracticable:

(i) random testing of the final product using a rabbit test.
(ii) a screen test for injectable solutions using millipore membrane filters.
(iii) random testing of the ingredients, the equipment and the processing technique.

Periodic pyrogen spot tests on batches of injections are useful to check the preparative technique in the hospital pharmacy. Where test facilities are limited it may be wise to employ them for:

(i) the solvent: for example, in the case of water for injections, when a still has been cleaned after a period of continuous use.
(ii) the solute: especially those in the high risk class. Batches of a given solute from different manufacturers may yield qualitatively varying febrile responses when tested for pyrogenicity.
(iii) equipment and apparatus: a trial run, using apyrogenic water or saline, will check that this has been correctly prepared and stored. Incorrectly cleaned and prepared apparatus may become pyrogenic within a few hours.
(iv) processing: random pyrogen testing of a prepared infusion will ensure that a suitable technique is being maintained in the hospital pharmacy or, if necessary, in the industrial laboratory for the preparation of injections.

Table I shows the results of some typical random tests for pyrogens carried out on infusion fluids. The positive response in sample 5 was due to incorrect preparation of the rubber caps of the bottles.

It is often more practical to use the screening test for routine pyrogen testing of injections than the pharmacopoeial test. This is particularly so in hospitals because the cost of pyrogen testing each batch of the increasing number of injections being prepared would increase the price. Marcus,

Anselmo and Perkins (1958) and Marcus, Anselmo and Luke (1960) found a relationship between the bacterial count of a solution and its pyrogenicity. Later Marcus (1964) described bacterial membrane filtration as a method of determining pyrogenicity. He found that a count of 10 organisms per millilitre indicated pyrogenic contamination in a freshly prepared solution. This represents a "safety factor" of about 100 to allow for non-viable pyrogenic sources in the material tested. Analysis of the bacterial flora from distilled water has shown that potent pyrogen-producing species of the Enterobacteriaceae are rarely found. Benfante and Labarre (1969) have

TABLE I

USE OF PERIODIC PYROGEN TESTING FOR THE CONFIRMATION OF SATISFACTORY MATERIALS AND PROCESSING METHODS FOR INFUSION FLUIDS

Sample	*Treatment*	*Average response in rabbits (°C)*
(1)	0·9% saline standard	0·07
(2)	Saline from Manesty still after reconditioning	0·33
(3)	Saline from Manesty still after reconditioning	0·44
(4)	Saline from Manesty still after reconditioning	0·32
(5)	Saline from Manesty still after reconditioning	0·72

shown that the screen test for pyrogens ensures that parenteral products conform to established standards of quality control provided that the culture assay is performed as soon as the product is prepared for sterilization. It is, however, a secondary type of quantitative procedure based on a bioassay in rabbits, and is consequently not completely infallible. Where pyrogenicity is suspected, after a screen test, the pharmacopoeial test should be carried out. Once a normal range of bacterial counts has been established, the adequacy of the preparative technique, the solvent(s) and the equipment and apparatus can be determined. A significant increase in the deviation of the bacterial count from an expected number will indicate the existence of an actual or potential source of pyrogens.

So far, I have been considering pyrogenic contamination of parenteral solutions. But even if a product has passed a pyrogen test, a pyrogenic reaction might still occur in a patient owing to a contaminated giving-set or an improperly prepared syringe. Using contaminated syringes I have found that the same number of organisms are necessary to yield a pyrogenic response as did Marcus. Of the four organisms tested, using a dose of 10^6 dead cells dried on to each syringe, only *Escherichia coli* yielded a pyrogenic response. *Bacillus subtilis*, in the same dose, did not give rise to a pyrogenic reaction, whereas *Staphylococucs epidermidis* yielded a greater response from an all-glass syringe than it did from a plastic one, probably indicating that the pyrogenic substances are more strongly adsorbed onto the plastic. A haemolytic *Streptococcus* species gave almost no response (see Table II). With this amount of contamination, a "safety factor" of about 100 is used for each syringe.

In practice, microbial contamination of water is due to a variety of organisms. Water from different localities contains different species; Leicester tap-water is contaminated by coliform organisms, a few small

TABLE II

PYROGENIC RESPONSES FROM SYRINGES CONTAMINATED WITH MICROORGANISMS AND THE SOLUTION ADMINISTERED IMMEDIATELY AFTER THE ADDITION OF SOLVENT

	Response (°C)		
	Syringe		
10^6 Microorganisms	*Plastic*	*All-glass*	*Saline control*
Staphylococcus epidermidis	0·30	0·46	0·38
Bacillus subtilis	0·40	0·21	
Escherichia coli	0·62	0·42	0·28
Streptococcus faecalis	0·13	0·26	
5×10^6 Microorganisms			
Escherichia coli	1·17		

cocci, some actively motile bacilli, probably of the pseudomonas group, and filamentous bacilli, whereas London tap-water contains aerobic spore-bearing organisms, gram-negative coliform organisms and small motile organisms, probably again belonging to the genus *Pseudomonas*. This contamination of tap-waters from different areas with differing species leads to varied pyrogenic responses. However, unless traces of tap-water are left in the syringe for long periods before sterilization, the contamination will not be sufficient to cause a pyrogenic response. Palmer (1967, pp. 103–126) has shown that the pyrogenicity of some, but not all, tap-waters is destroyed by autoclaving. As with purified pyrogens, in no circumstance may a sterile solution be assumed to be apyrogenic. This presents a problem when using the screen tests for evaluating pyrogenic action, for only viable organisms are cultured, whereas pyrogens may also have been produced by non-viable organisms.

Sterilization using the pharmacopoeial ionizing radiation dose is inadequate to destroy the pyrogens present. Hutchinson and Whittet (1957) found that a dose of $3 \cdot 0 \times 10^6$ rads (rd) had little effect on the pyrogenic content. With London tap-water, a dose of at least $5 \cdot 0 \times 10^6$ rd was necessary before the sample passed the B.P. test, and with the purified pyrogen, Pyrexal, the dose required was 25×10^6 rd.

Recently, irradiation of syringes contaminated with exo- and endo-pyrogens from 5×10^5 organisms of *Escherichia coli* was studied. When solvent was left for 10 minutes inside a syringe sterilized with $2 \cdot 5 \times 10^6$ rd, before being injected, pyrogenicity had not been removed (see Table III). All products and equipment used for parenteral administration must therefore be sterilized as soon as possible after a correct preparation, when there is little likelihood of syringes being sufficiently heavily contaminated to render the injection pyrogenic.

TABLE III

PYROGENIC RESPONSES FROM SYRINGES CONTAMINATED WITH 5×10^6 ORGANISMS OF *Escherichia coli*

	Endopyrogen (°C)	*Exopyrogen* (°C)	*Control* (°C)
Plastic syringe			
0 min	0·58	0·46	
10 min storage	1·50	1·25	0·44
Glass syringe			
0 min	0·58	0·48	
10 min after irradiation	0·58	0·62	0·43

Certain pharmaceutical products, such as medicaments which are unstable in solution and radiopharmaceuticals, present special problems when being tested for pyrogenicity. With the medicaments, delay in performing a pyrogen test may be sufficient to render the product unsatisfactory for use, therefore a pyrogen test should be carried out on the solute(s) and solvent independently by the hospital or industrial laboratory. Greater problems are encountered when radiopharmaceuticals are tested because in addition to any pyrexia caused by the pyrogens present, there is also the possibility that the radioisotope itself may exert a febrile action. The larger the absorbed dose of radiation, the greater will be any febrile effect. Errors may therefore result in the testing of such products by the pharmacopoeial test. Charlton (1965) suggests that non-testing of radiopharmaceuticals before despatch to the user is acceptable for two reasons:

(i) there may be serious radioactive decay during the period of testing.
(ii) the radioactivity might interfere with the test.

In industry a complete day must be allowed for a pyrogen test, particularly if the first test is equivocal. When this period is added to the time necessary for the preparation, packaging and transport of the isotope, the time factor is very significant because many medicinal investigations are carried out with short-lived isotopes at the beginning of a week.

In the pyrogen test, the volume of a product administered to each rabbit should be related to the volume normally given to a human. For radiopharmaceuticals this volume is small, and for a therapeutic radioactive product, the radioactivity associated with this volume may be very large. Storage of the product before carrying out a pyrogen test will almost certainly be necessary to allow sufficient decay of the radioactivity. For example, the *British Pharmacopoeia* 1964 Addendum specifies storage of ^{198}Au as colloidal gold, initially 10 millicuries, until the activity is 100 microcuries. Diagnostic radioactive injections usually contain less than 100 microcuries of radioactivity, making storage before pyrogen testing unnecessary.

Because the radiopharmaceuticals are of small volume, the quantity which can be tested for pyrogens is extremely small. A realistic test dose is

TABLE IV

PYROGEN TESTS ON ANTI-HAEMOPHILIC GLOBULIN SAMPLES USING DIFFERENT METHODS OF PREPARATION

Sample	*Result* (°C)
(1) Porcine	1·38
(2) Porcine	1·48
(3) Porcine	0·70
(4) Bovine	1·26
(5) Bovine	0·18
(6) Solvent (control)	0·21

0·1 of the therapeutic dose per kilogramme body weight of the rabbit. This dose was recommended in Germany (Burianek, 1965) and in the U.S.A. However, a positive pyrogen response will only be obtained from very heavily contaminated samples.

Of increasing commercial interest are the elution column generators for very short-life radioactive isotopes which provide a source of radioactive material over a period of days. Consequently there is a strong likelihood of microbial contamination of the column, and also, since the column adsorbent may be a medium suitable for bacterial growth, the eluates may be pyrogenic. The generator column should be kept in as aseptic conditions as possible to keep the bacterial flora as low as possible and so ensure minimal pyrogenic content.

Industrially, the production of apyrogenic solutes is essential where the medicament is a good medium for initiating and maintaining bacterial growth. Pyrogen testing during the manufacture of such substances will establish the best process for preparing them (see Table IV) and ensure that the resulting solute is apyrogenic. Manufacturers producing medicaments by different methods are likely to obtain substances with differing pyrogenic actions. Solutes for injections should therefore be chosen for their relative apyrogenicity and not necessarily for the cheapness or chemical purity of the materials (see Table V).

Large amounts of apyrogenic solvents are required in both hospitals and industry. Water for injections is commonly needed and is required by the *British Pharmacopoeia* to be produced by suitable distillation. Commercial stills used for this purpose include the Manesty Automatic, the Mascarini Thermocompressor, the Alembic, the Barstead, and the Glastechnik Single Distillation Unit. Entrainment of droplets containing non-volatile pyrogens is prevented by baffles or columns, so ensuring an apyrogenic condensate.

TABLE V

COMPARISON OF RESPONSES TO DIFFERENT SOURCES OF MANNITOL AND SODIUM CITRATE

Substance	*Response* (°C)
(1) Mannitol	0·04
(2) Mannitol	0·29
(3) Mannitol	0·53
(1) Citrate	1·63
(2) Citrate	0·78
(3) Citrate	1·08
(4) Citrate	0·82
(5) Citrate	0·37

From the many tap-waters tested, I have found only that from Stoke-on-Trent to be apyrogenic. It must never be assumed that subsequent sterilization of the products will always remove the pyrogenicity. Equally, it must never be assumed that a sterile sample of water for injections is apyrogenic, because pyrogens remain in the solvent after the death of the micro-organisms. Routine periodic pyrogen tests on the solvent should be carried out, and when a still has been dismantled to remove mineral deposits and for cleaning, the first distillate should be tested for pyrogen before the solvent is used for preparing injections, to ensure that the water is of a suitable quality to be used in manufacturing products.

Purified water is not as yet accepted by the *British Pharmacopoeia* as being of a suitable standard for the preparation of injections. However, Saunders, Lorch and Hassell (1969) have shown that combined with macroreticular resin as a filter, ion-exchange resins remove purified pyrogens from solution. Whittet (1956, 1958) and Palmer (1967, pp. 154–179) have shown that ion-exchange resins remove the pyrogenicity from tap-water, and it has also been shown that ion-exchange resins remove some or all of the pyrogenicity from solutions of purified pyrogens (Palmer, 1967; Palmer and Whittet, 1971). The approach in Britain is more cautious than that on the continent, where use of water from ion-exchange commercial mixed-bed resin columns has been advocated for injections. It is possible that ion-exchange resins, combined with filtration, may be adopted in the near future for preparing water for injections. If this is so, adequate quantities of solvent will be available in hospitals and industry for the manufacture of injections, and thus delays due to scarcity of apyrogenic solvent would be avoided.

Strict precautions will always be necessary for preparing injections. Although scientific progress has given much information about pyrogenic contamination, pyrogens will remain a hazard wherever microorganisms are found.

CONCLUSIONS AND SUMMARY

(1) Infusion fluids and other injections containing nutrient medicaments are the preparations most likely to be contaminated by pyrogens. Periodic testing of these preparations will confirm that a rigid technique has been used throughout their production.

(2) Checks on: (i) solute(s), (ii) solvent(s), (iii) equipment and apparatus, and (iv) processing, will ensure that the final products are satisfactory.

(3) Where adequate facilities do not exist for pyrogen testing, provided (1) and (2) are carried out, pyrogen tests on each batch of injections prepared in hospitals and industry are unnecessary.

(4) Membrane filtration is a good screening test for pyrogens in injectable solutions. Where excessive microbial contamination is found, the absence of pyrogens should be confirmed by a rabbit test.

(5) Contamination of medical equipment and apparatus, such as syringes, by at least 10^6 microorganisms is extremely unlikely. If the injections have been properly prepared, pyrogenic reactions due to contaminated equipment will be very rare in patients.

(6) Radiopharmaceuticals, and medicaments which are unstable in solution, present special problems in pyrogen testing.

(7) In industry, production of apyrogenic solutes and solvents is essential. Pyrogen testing can be used to establish the best process for manufacture, and the most suitable solute and solvent to use for the preparation of pharmaceutical products.

(8) The use of ion-exchange resins, in combination with filtration, may become an accepted method for producing water for injections.

REFERENCES

BENFANTE, P., and LABARRE, J. (1969). *Drug Intelligence Clin. Pharmac.* **3,** 286.
BENNETT, I. L., WAGNER, R. R., and LEQUIRE, V. S. (1949*a*). *J. exp. Med.*, **90,** 335–347.
BENNETT, I. L., Wagner, R. R., and LEQUIRE, V. S. (1949*b*). *Proc. Soc. exp. Biol. Med.*, **71,** 132–133.
BILLROTH, T. (1865). *Arch. klin. Chir.*, **6,** 372–495.
British Pharmacopoeia (1968). pp. 1348–1349. London: Pharmaceutical Press.
BURDON-SANDERSON, J. (1876). *Practitioner*, **16,** 257–337, 417.
BURIANEK, J. (1965). *World Health Organisation*, Pharm. Ed. Sect., No. 122, pp. 2–3.
CENTANNI, E. (1894). *Chem. ZentBl.*, **6,** 597.
CHARLTON, J. C. (1965). *World Health Organisation*, Pharm. Ed. Sect., No. 123, pp. 1–2.
CO TUI and SCHRIFT, M. H. (1942). *J. Lab. clin. Med.*, **27,** 569–576.
HORT, E. C., and PENFOLD, W. J. (1911). *Br. med. J.*, **2,** 1589–1591.
HORT, E. C., and PENFOLD, W. J. (1912). *Proc. R. Soc. Med., Path Sect.*, **5,** 131–139.
HUTCHINSON, J. P., and WHITTET, T. D. (1957). *J. Pharm. Pharmac.*, **9,** 950–954.
International Pharmacopoeia (1955). 1st edn, vol. II, pp. 283–284. Geneva: WHO.
JONA, J. L. (1916). *J. Hyg., Camb.*, **15,** 169–194.
MARCUS, S. (1964). *Bull. parent. Drug Ass.*, **18,** 18–24.
MARCUS, S., ANSELMO, C., and LUKE, J. (1960). *J. Am. pharm. Ass., Sci. Edit.*, **49,** 616–619.
MARCUS, S., ANSELMO, C., and PERKINS, J. J. (1958). *Proc. Soc. exp. Biol. Med.*, **99,** 359–362.
PALMER, C. H. R. (1967). Ph.D. Thesis. London University.
PALMER, C. H. R., and WHITTET, T. D. (1961). *J. Pharm. Pharmac.*, **13,** 62T–66T, suppl.
PALMER, C. H. R., and WHITTET, T. D. (1971). *Chem. Ind.*, in press.
SAUNDERS, L., LORCH, W., and HASSELL, M. H. (1969). *Soc. Chem. Ind.*, 106–110.
SEIBERT, F. B. (1923). *Am. J. Physiol.*, **67,** 90–104.
United States Pharmacopoeia (1955). 15th revision, pp. 883–884. Easton: Mack.
WAGNER, R. R., and BENNETT, I. L. (1950). *J. exp. Med.*, **91,** 135–145.
WHITTET, T. D. (1956). *J. Pharm. Pharmac.*, **8,** 1034–1041.
WHITTET, T. D. (1958). Ph.D. Thesis, pp. 100–132. London University.
WYLIE, D. W., and TODD, J. P. (1948). *Q. Jl Pharm. Pharmac.*, **21,** 240–252.
WYLIE, D. W., and TODD, J. P. (1949). *J. Pharm. Pharmac.*, **1,** 818–835.

DISCUSSION

Grundman: Did you do any experiments showing that the temperatures of the rabbits rose between 11·0 a.m. and 2·0 p.m.? We found that the temperature rises occasionally, sometimes more than 0·3°C, but that the mean temperatures of a lot of rabbits remains level throughout the day (unpublished).

Palmer: We saw this rise in temperature whether the animals were treated or not.

Whittet: The temperature peak was at about 2·0 p.m. so we nearly always injected the animals after 2·0 p.m., and at the same time each day (Whittet, 1958). When doing a British Pharmacopoeia test one measures the highest temperature reached and if there is a rise for only a short period during the whole three hours, then that is recorded as the highest one.

Cranston: These materials are being given to patients so in fact pyrogen testing is going on all the time. This is not the ideal way to control manufacture, but at least it provides a warning system.

Myers: Villablanca and I (1965) found that *Salmonella typhosa* injected into the cerebral ventricles in a quantity as little as 10^3 organisms/ml, caused a fever. This test is 1000-fold more sensitive than yours. We have also obtained a pyrexic response in the cat by microinjecting as little as 500 organisms into the hypothalamus. So intraventricular or intracerebral injections in the rabbit may be a more suitable and accurate test for pyrogenicity.

Palmer: They probably would be, but you are going to run into practical testing problems.

Cooper: Pyrogenicity is only one aspect of responses to material derived from microorganisms. Does the pyrogenicity represent a reasonably parallel assay of the other possible toxic side-effects of these organisms?

Palmer: I have no information about any parallelism. The rabbit is accepted as being more sensitive than a normal human, and the dose administered is very large compared with a human dose (infusions of 10 ml/kg).

Smith: Since so much transfusion fluid is used in this country without being pyrogen-tested and without adverse reactions, patients are perhaps very tolerant towards pyrogens. I have been engaged in controlling the pyrogenicity of transfusion solutions from large-scale production. As part of our "good housekeeping" we demand that the preparations are autoclaved immediately after manufacture, or certainly within four hours. All batches are submitted to pyrogen tests and we have had no difficulty in producing pyrogen-free distilled water and normal salines on this large scale. We test every batch of transfusion fluid which is prepared, but not necessarily to the full pharmacopoeial tests. The pyrogen test of the British Pharmacopoeia is based on a sequential sampling plan. We therefore inject two rabbits for each batch and do not pursue the examination even if the demand of the plan for two rabbits is not met, providing that the responses of both rabbits are below 0·6°C and the demand of the plan of treating a whole day's production as one batch is met.

About 14 years ago we had to reject quite a large number of glucose salines for pyrogenicity because a particular batch of glucose was at fault. More recently, we have had similar experiences with fructose. In large scale production where personnel may change suddenly the only safe way is to test every batch of prepared solution. If one adopted random sampling and found a particular batch to be pyrogenic how much retrospective testing would one need to do? A batch is 500 litres, which rules out the possibility of using columns to prepare this on an industrial scale.

Palmer: An ion-exchange column can be switched on first thing in the morning and the water taken out, thus saving time.

Smith: But your column, Dr Palmer, is only about 10 litres.

Saunders: The threat of pyrogens has always held up the use of ion exchange columns for producing water for injection. Recently a new type of resin with a lot of internal surface has been developed. With the gel-type resin these big pyrogen molecules are in contact with the active groups on the resin only on the surface of the particles, but with these macroreticular resins interactions occur right through the resin particles; consequently you can put up flow speeds and reduce the size of the columns, and the columns can also be sterilized by gamma irradiation.

Work: What guarantee have you that your ion-exchange column has remained pyrogen-free during the time you were not using it?

Saunders: The equipment has to be matched to your requirements so that it is only used over a very limited period.

Work: Do you sterilize it every time before you use it?

Saunders: After use the sterilized ion-exchange column will be replaced by another; it is proposed to make these columns on a disposable basis.

Bondy: In clinical work the biggest source of pyrogens is the apparatus that is used to administer the fluids rather than the fluids themselves, and since we started using entirely disposable equipment this problem has disappeared. Although the number of organisms in a litre of saline would be very large, the number of organisms in the socket of a needle could be immense. Since the size of the pyrogen is very different from the size of the radioactive material that Dr Palmer is interested in, a simple way of separating the pyrogens from the radioactive material might be by using a molecular sieving system.

Whittet: I supervised the preparations of injectable preparations in a hospital pharmacy for 20 years. I investigated well over 100 samples where a pyrogenic reaction was suspected and where a good technique had been used I never found any pyrogen. Most often it was present because a nurse on the ward had opened a bottle and used it later; once a tray of bottles did not get in the autoclave, and these were very pyrogenic. I agree with Mr Smith that as far as possible routine tests should be done, at least on the commonly used substances, and I hope that we may be able to do this on a regional scale in the hospitals eventually.

Another type of hazardous injection is of intraperitoneal solutions—these must be pyrogen-free. One of the most terrific pyrogen reactions I came across was with a sample of radioactive colloidal gold. I tested it and was only able to use 0·3 ml/kg instead of the 10 ml/kg of the BP test. I asked what method of sterilization had been used, and was told none; it had been assumed that the irradiation would sterilize it. This is a problem with the columns which Dr Palmer mentioned; unless the physicist involves the pharmacist, who is aware of pyrogens, very severe reactions can be caused.

REFERENCES

VILLABLANCA, J., and MYERS, R. D. (1965). *Am. J. Physiol.*, **208**, 703–707.
WHITTET, T. D. (1958). Ph.D. Thesis, University of London.

THE DILEMMA OF QUANTITATION IN THE TEST FOR PYROGENS

D. R. BANGHAM

National Institute for Medical Research, London

THE Pyrogen Test is an important quality control test for preparations administered parenterally, and it is essential for the protection of the patient. The adverse effect of a pyrogen reaction is unpleasant for anyone; it is particularly hazardous to a patient who is critically ill or undergoing major surgery, when it may be the (unrecognized) cause of acute collapse. Patients repeatedly given a drug contaminated with a pyrogen, by any parenteral route, are in danger of becoming acutely sensitized to it as an antigen. Yet it is salutary to remember that this official test relies entirely on the "absolute" sensitivity (or insensitivity) of the particular rabbits used to the pyrogen(s) that may be contaminating the preparation under test. Every effort to make the test into a practical and generally applicable comparative bioassay —in terms of a common reference material or standard—has failed. Why?

There are two reasons. One is because of the heterogeneity of bacterial pyrogen molecules encountered in nature; the other is because of the variability of test animals.

The validity of comparative bioassays depends on an assumption that like is compared with like; that the test substance behaves in the test system in a way identical to the standard used to the extent that it behaves like a dilution of the latter. Pyrogen tests cannot in practice comply with this requirement because of the strain-diversity of individual pyrogens within the general class of bacterial pyrogen lipopolysaccharides. This non-comparability is a recurrent and as yet unsolved problem in the bioassay of heterogeneous bacterial products, even when one end-product is selectively and intentionally isolated (such as purified protein derivative from different strains of tubercle bacilli). Moreover, in testing a product for chance contamination it is not known which individual pyrogens may be present, or in what proportions: it is thus quite impossible to anticipate (let alone provide) what would be an appropriate pyrogen standard to assay it with.

A second reason is because of the unknown and unpredictable differences in sensitivity of each test animal to pyrogens or mixtures of them. Apart from the intrinsic variability of individual animals, even from an inbred strain, the sensitivity of each varies with its degree of "tolerance" or "immunity" to different pyrogens.

Tolerance or immunity or sensitization to a given pyrogen may be pharmacological or immunological (e.g. Watson and Kim, 1964); and the nature of the mechanisms and the extent to which each is involved have been discussed at this symposium. We have heard to what extent the responsiveness of an animal to a given pyrogen is influenced by, among other things, its previous exposure to that and to other pyrogen molecules. Since the ways in which any individual test rabbit encounters various bacterial endotoxins are many, including ingestion, inhalation and injection, its sensitivity may be unpredictable and uncontrolled.

These general facts have been established in the literature for many years and formed the background for the formulation of the official test some 15 years ago. There has been little significant change since, for example, the test described in the *British Pharmacopoeia* (1958). Essentially the same test is officially prescribed for regulations for manufacture by the World Health Organization, Therapeutic Substances Act of Great Britain, Food and Drug Administration, National Institutes of Health of the U.S.A., and also in most pharmacopoeias. (In Britain tests described in the Pharmacopoeia are those expected to be applied only to the end product.)

I shall here outline some of the general principles underlying the formulation of the current test for pyrogen, and draw attention to some particular current problems. My purpose is to call attention to the weaknesses of the current test procedure and to ask that it be re-examined in the light of the latest knowledge of pyrogens and their mechanism of action.

PRINCIPLES UNDERLYING THE PRESENT PYROGEN TEST

The test is not, and has never been intended to be, a guarantee of freedom from pyrogen. Like the test for sterility it is a check against gross contamination, and the statistical chances of detecting contamination are a deliberate compromise between unworkably extensive testing and a reasonably useful test for safety. The test is not a quantitative assay in terms of a reference material. Nor is it possible officially to specify "the level of sensitivity" of the test animals. Some laboratories do a "test for sensitivity" of their own, but the limitations and shortcomings of this are discussed later.

Since it is impossible to quantitate pyrogens on an absolute scale the most that can be done is to prescribe in some detail the conditions in which the test is carried out. The influence of stress in suppressing a response to a pyrogen ("emotional hypothermia") was early recognized. The test is therefore carried out in a quiet uniform environment with a minimum of external stimulation. Uniform conditions of temperature and lighting are maintained and there should be a minimum of noise, handling or movement. Rabbits are used since their body temperature under test conditions is steadier than that of most other small laboratory animals, and they can be trained to remain quiet for several hours with a minimum of constraint. Their large size makes them convenient for handling for the intravenous administration of test doses into an ear vein. They are trained for two or three weeks beforehand to sit quietly in stalls, and to accept the injection of a saline solution as a control test and the insertion of thermometer or recording probe without being upset. Animals selected for the test are those which, according to specified criteria, have not previously shown an undue rise in temperature after the intravenous administration of pyrogen-free saline. Those whose temperature has been raised above a certain limit are excluded altogether from use in a test, whilst those that have shown a little rise in temperature may be re-used after a resting period of three weeks. Those which have received a dose of a substance which is obviously potentially antigenic should not be tested again with any substance that may reasonably be expected to cross-react immunologically with it.

The dose per kilogramme of the body weight of the rabbit is specified for each preparation. Generally it varies (among national requirements and pharmacopoeias) between the amount which would be regarded as a single normal human dose, and one-tenth of that amount. The basis on which the amount is decided seems to be sometimes somewhat arbitrary. It was certainly influenced by, if not derived from, Dare's observations on the relative amounts of a dose of *one* preparation of a crude extract of *Proteus vulgaris* (Dare, 1953) which gave an incidence of positive responses in his rabbits comparable to the incidence of pyrogenic reactions in human volunteers (Dare and Mogey, 1954). In certain instances the dose specified for the test may be restricted by the toxicity of the drug itself. Sometimes it is based on manufacturers' experience of what routine production batches of a product would pass.

After establishing a control baseline for an hour or so before the dose, the temperature is recorded at 30-minute intervals for three hours. The criteria for passing or failing a batch are applied to the figures for the *maximum* increase in temperature above the previous baseline temperature. This maximum may occur at any one of the recorded times in the following three hours; the integrated area under the temperature plot is not used as the parameter. If it is the maximum figure that is relevant then a more certain way to detect it would be with a continuous temperature record, such as can be obtained with modern electrical devices, rather than by relying on the intermittent observations at present stipulated.

I do not intend here to discuss the numerical criteria for passing or failing since these are spelled out in the official specifications. If a preparation fails (within limits) it is allowed to be retested on two or three additional groups of three animals. The selection of the best method for evaluating the results, by a simple additive procedure (as in the *United States Pharmacopeia*) or by K. L. Smith's cumulative procedure (which the 1968 *British Pharmacopoeia* prefers), is a statistical nicety.

The criteria for passing or failing a batch present no problems if there is no pyrogen in the preparation or if it is quite obviously contaminated. The difficulty comes in the elaboration of rational criteria for decision on borderline responses where it is difficult to reconcile the rigid numerical specifications of the present approach with all that is known of the intrinsic variability of the response in rabbits and in man.

If the rationale of the scale of sensitivity of test is so tenuous, how well does it work in practice? Fortunately the great majority of preparations show no pyrogenic effects in patients. (This is most certainly attributable to good pharmaceutical manufacturing practice rather than to the reliability of the test.) Decision in borderline cases may mean large financial implications for manufacturers and presents ethical problems in research where preparations of rare materials, such as human hormones, are involved. However, there are some preparations which do cause troublesome reactions in man and which on retesting in rabbits may not necessarily fail the test. (Some of those that do fail on retest may contain those fortunately rare slow-growing bacteria which also fail to show up in the seven- or ten-day sterility test.)

The current test thus cannot be regarded as completely reliable and attempts to improve it have been made over the years. Starting in 1952 a large international collaborative study was carried out to see whether ampouled preparations of a partially purified extract of *Proteus vulgaris* and a purified lipopolysaccharide from *Serratia marcescens* could be assayed in terms of each other. The great amount of data collected showed how consistently a reproducible potency ratio between the two preparations could *not* be obtained. Moreover the more dose levels used, the more bizarre the shape and slope of the log-dose response relationship. The results were not suitable to report in more than summary form and attempts to devise a bioassay ceased. Nevertheless it was thought that the provision of a common reference material might facilitate research work on quantitation and mechanism of action; on these grounds the International Reference

Preparation for Pyrogens was established by the World Health Organization (Humphrey and Bangham, 1959). This consists of a freeze-dried preparation of the O somatic antigen of *Shigella dysenteriae* (Shiga) made and characterized by Davies, Morgan and Mosiman (1954).

Recently there have been renewed attempts to test or standardize the sensitivity of rabbits used in the pyrogen test. In 1968 we asked about fifty national control and manufacturers' laboratories what, if any, steps they took in selecting rabbits. More than thirty tested their animals either routinely every few months or on entry into the pyrogen colony with one preparation or another. Each laboratory had chosen for itself a test material, the dose level, and the response criterion for accepting or rejecting the animal. In some laboratories 10–15 per cent of rabbits are turned down as unsuitable on grounds of such a test. The International Reference Preparation itself was the most popular test material; the dose used for this "test for sensitivity" varied. One laboratory used 0·0015 μg/kg, most (seven) used 0·002–0·005 μg/kg, and one laboratory used 0·5 and even 10·0 μg/kg of rabbit body weight. Most laboratories used the International Reference Preparation at a dose of 0·003 μg/kg. This happens to be the quantity stated in the memorandum on the International Reference Preparation as the dose that raised the temperature of some rabbits in the laboratory where the preparation was made.

The most constructive interpretation one can make of those replies is that they show how much people want to make the test quantitative.

I do not think there is much need to make the test more sensitive. There *is* a need for means to increase the uniformity of the test animals if reliance on the absolute response of animals is inescapable. For example, would young rabbits be more uniform even though they may be less sensitive?

Besides their nuisance value in reasonably healthy patients, there are situations where contaminated materials present especial hazards. Thus pyrogen reaction in patients undergoing major surgery with extracorporeal circulation, or in those on renal dialysis or otherwise severely ill, may be disastrous. Certain drugs are isolated from materials which carry a high intrinsic risk of contamination with pyrogen which is not associated with contamination during manufacture. Products of microbial fermentation, such as antibiotics, dextrans and streptokinase, and hormones and enzymes isolated from human urine are examples.

What is the clinical risk of a Shwartzman reaction in a patient given one of a pyrogen-contaminated product daily or weekly for a long period? What additional safeguards could possibly be applied to drugs given intrathecally? How does irradiation of test rabbits with radioactive isotopes of various emission spectra influence the pyrogen response?

A better understanding of the body's physiological response to bacterial pyrogens may help to answer these pressing and pragmatic questions. It may even suggest how to devise a bacterial pyrogen assay.

REFERENCES

British Pharmacopoeia (1958). Pp. 947–948. London: Pharmaceutical Press.
British Pharmacopoeia (1968). Pp. 1348–1349. London: Pharmaceutical Press.
DARE, J. G. (1953). *J. Pharm. Pharmacol.*, **5**, 528–546.
DARE, J. G., and MOGEY, G. A. (1954). *J. Pharm. Pharmacol.*, **5**, 325–332.
DAVIES, D. A. L., MORGAN, W. T. J., and MOSIMAN, W. (1954). *J. Biochem., Tokyo*, **56**, 572–581.
HUMPHREY, J. H., and BANGHAM, D. R. (1959). *Bull. Wld Hlth Org.*, **20**, 1241–1244.
United States Pharmacopeia XVII (1970). Pp. 863–864. Easton: Mack.
WATSON, D. W., and KIM, Y. B. (1964). In *Bacterial Endotoxins*, pp. 522–545, ed. Landy, M., and Braun, W. New Brunswick: Rutgers University Press.

DISCUSSION

Cranston: I can't see that the possible use of leucocyte pyrogen that you suggested (p. 78) would solve the pharmaceutical difficulties. It wouldn't solve the important problem of previous exposure to endotoxin, so it would only be usable as a standard for the rabbit response, irrespective of its previous exposure to bacterial pyrogen.

Bangham: I was only suggesting that this was an aid to research work.

Myers: If one is interested in whether there is a linear relationship between the pyrogenicity of different organisms one needs an extremely sensitive assay preparation. For research purposes, I think the intraventricular or intrahypothalamic routes are probably the most sensitive, but these routes could not be used for standard pyrogenicity testing.

Cranston: There could be a case for using one of these routes when testing products intended to be injected intrathecally. In animals there is evidence that the sensitivity to pyrogen given intrathecally is considerably greater than to pyrogen given intravenously (Bennett, Petersdorf and Keene, 1967; Adler and Joy, 1965).

Snell: The effect of reticuloendothelial blockade with thorotrast is terribly complicated. It doesn't really abolish tolerance; it increases the sensitivity to bacterial pyrogen enormously so that it looks as though you have abolished tolerance. It probably affects opsonic factors in blood that are partly responsible for clearing bacterial pyrogen as well as the cells of the reticuloendothelial system.

Cooper: The important point is not that it abolishes tolerance, but that it reduces the variability of results.

Landy: The reticuloendothelial blockage produced by Beeson (1947) with Indian ink is now recognized as due largely to the presence of bacterial endotoxin in the ink.

Smith: Rabbits do vary from colony to colony and possibly within colonies, but workers in different parts of the world, producing a routine test of freedom from pyrogens, hit on the same sort of dose. This must have meant that there was some uniformity between rabbits. When we were designing the pharmacological test, we had evidence from Dare (1953) that the variance of a single responding rabbit was about 0·127 and this compared well with the figure of 0·11 that I had from my own laboratory. And Dr Murphy (1971) found that the residual variance of a single response was again about 0·11, which seems to illustrate a great degree of uniformity of rabbits within any one colony.

Landy: The figure would differ greatly depending on whether one had used bacterial pyrogen or endogenous pyrogen, because the act of administering bacterial pyrogen alters the animal's reactivity.

Smith: The residual variance measures the uniformity of rabbits receiving the same preparation within the same laboratory.

Snell: Incubation of white cells with bacterial pyrogen not only generates leucocyte pyrogen, but also does other things to the white cell; the permeability increases, and enzymes such as aldolase come out. This might form an *in vitro* biochemical assay system, and we could get rid of the rabbit altogether.

Pickering: Is this peculiar to your endotoxin or does it happen with all bacterial pyrogens?

Snell: I don't know about the pyrogens from gram-positive bacteria, but I should imagine it would happen with phagocytosis of those.

Cranston: You would probably get rid of a lot of the variability if you used your test material to make leucocyte pyrogen and then tested the leucocyte pyrogen by bioassay.

Snell: Then you are back to the problem of a bioassay of a pyrogen.

Cranston: Bioassay of leucocyte pyrogen would have fewer of the problems which result from the animals' previous experience of bacterial pyrogen.

Bodel: It is also a very sensitive way to demonstrate endotoxin, as has been shown recently by Moore and co-workers (1970) in a system using exudate cells, as well as by Dr Cooper (1971).

Saunders: Is there any evidence that the integrated temperature-time curve will give more reliable results than the maximum temperature?

Cooper: We used to use the area beneath the rise and fall in the temperature curve, but Dr Murphy showed that the group results don't fall on a normal distribution curve, and the skewness of the distribution interferes with the statistical calculations.

Pickering: Is the absolute temperature rise any better as far as distribution is concerned?

Cooper: Yes, these measurements fall on a normal Gaussian distribution.

REFERENCES

Adler, R. D., and Joy, R. J. T. (1965). *Proc. soc. exp. Biol. Med.*, **119**, 660–663.
Beeson, P. B. (1947). *J. exp. Med.*, **86**, 39–44.
Bennett, I. L., Petersdorf, R. G., and Keene, W. R. (1967). *Trans. Ass. Am. Physns*, **70**, 64.
Cooper, K. E. (1971). This volume, pp. 5–17.
Dare, J. G. (1953). *J. Pharm. Pharmac.*, **5**, 898.
Moore, D. M., Cheuk, S. F., Morton, J. D., Berlin, R. D., and Wood, W. B. (1970). *J. exp. Med.*, **1131**, 179.
Murphy, P. A. (1971). This volume, pp. 59–73.

GENERAL DISCUSSION

THERMOREGULATION

Pickering: If the mechanism that regulates body temperature has the classical three parts—a receptor mechanism that appreciates changes of temperature, a central coordinating mechanism, and an effector mechanism —where is the receptor site, and how are the hot and cold firing neurons involved?

Cooper: Undoubtedly there are two types of units in the hypothalamus, and also further back in the brain stem, which respond to local heating and cooling by changing their firing rates. But a clear correlation between the behaviour of these units and the thermoregulatory responses has not been demonstrated.

Pickering: Are the receptors confined to the places that we know from other evidence are concerned with temperature regulation?

Cooper: Their highest density is in the anterior hypothalamic pre-optic area, but they can be found as far back as the mid-brain, which agrees with the latest evidence on sites at which leucocyte pyrogen acts.

Pickering: But you have to postulate a receptor which appreciates changes in temperature and it isn't necessarily that receptor which is responsive to leucocyte pyrogen.

Cooper: I agree, but both appear to be in the same place. I suppose some of these temperature "receptors" are distributed more widely in the hypothalamus because the blood supply to the brain comes from two ends, and this gives you a double chance of picking up changes in arterial blood temperature.

Cranston: There is evidence that the anterior and posterior temperature-sensitive neurons are interrelated. Nakayama and Hardy (1969) could alter the firing rate of the front ones by heating and cooling the back ones, but not *vice versa*.

Myers: People continually forget that if the posterior hypothalamus is lesioned bilaterally and the area is destroyed dorsal to the mammillary bodies including the zona incerta and the fields of Forel, the animal remains afebrile when a bacterial pyrogen is injected intravenously.

Pickering: Can the animal regulate its body temperature?

Myers: No. In the monkey, for example, poikilothermia persists for 20 to 30 days. Only then may some recovery in thermoregulation occur (see Myers, 1971).

Pickering: Does the response to pyrogen come back?

Myers: That has never been tested to my knowledge.

Pickering: It ought to be, because if the animal loses the ability to regulate body temperature all that you know is that you have destroyed an essential part of the regulatory system, but not which part.

Is the mid-brain, the place where your salicylate works, Dr Rawlins, a place that responds to endogenous pyrogen?

Rawlins: It would appear to be, or the response may occur in a closely related area.

Cranston: Rosendorff, Mooney and Long (1970) have been getting responses from somewhere in the area of the mid-brain.

Myers: A pyrexia can be elicited by a pyrogen injected into that area, but the *latency*, which is a critical measure, is much longer. If the response to a pyrogen injected into the brain-stem does not occur for over 30 minutes, then the finding is the same as ours (Villablanca and Myers, 1965). Diffusion of these pyrogenic organisms apparently takes place in spite of their molecular weight. Eventually the organisms will reach the anterior regions of the hypothalamus where they exert their pyrexic action (Villablanca and Myers, 1965). In the cat and in the monkey, the mesencephalon seems to be unresponsive to microinjections of pyrogen. So we are again faced with this question of species differences, or perhaps the less awesome problem of differences in experimental technique.

Cranston: Rosendorff, Mooney and Long (1970) used leucocyte pyrogen, and not organisms. Are you talking about leucocyte pyrogen or bacterial pyrogen?

Myers: In the monkey bacterial pyrogen, and in the cat either.

Teddy: I injected leucocyte pyrogen into a similar site in the mid-brain in only 15 rabbits and got a rise of about 0·5°C after roughly three-quarters of an hour.

Pickering: Dr Feldberg has suggested that pyrogen affects the setting of the central mechanism by removing the "calcium brake". What is the "calcium brake"?

Myers: Veale and I have done some work which may not support Dr Feldberg's concept of the "brake". If the sodium and calcium concentrations do function to control the set-point, then they do so as a ratio (Myers and Veale, 1970, 1971). If the sodium and calcium concentrations are doubled in the hypothalamic perfusate, the temperature of the cat remains the same. If the concentrations in extracellular fluid are lowered, as we believe they are after sucrose has been perfused through the same area of the hypothalamus, again the temperature of the cat remains stable. The temperature changes only when the ratio between the two ions is altered.

Cranston: One can make sense of this from the current views of cell membrane physiology. But has this anything to do with the way in which pyrogen acts, or is it just a general phenomenon of altering cell membrane functions?

Pickering: Beeson (1938) used to give injections of calcium gluconate to patients with rigors, and he found that it was most effective; it stopped them shivering almost at once and quietened them down, but he thought it never made any difference to the fever.

Cranston: Again, this could be a peripheral effect due to a high calcium concentration.

Bondy: Dr Landy has pointed out that materials capable of producing fever are probably constantly available in tiny amounts. Perhaps reduction in the calcium concentration just lowers the threshold so that the pyrogens can be effective, whereas when enough calcium is around the threshold is high enough for the normal ambient pyrogens not to produce an effect.

Cooper: It has been shown (Hardy, Hellon and Sutherland, 1964; Wit and Wang, 1968) that the action of pyrogen on one type of temperature-sensitive neuron is to increase the sensitivity, while on the other type of neuron it is to decrease it. I would have thought that the effects of sodium and calcium on nerve fibres would be constant in all types of nerve unit.

Myers: This is why our so-called sodium-calcium set-point couldn't possibly be in the anterior hypothalamus. Anatomically, the posterior area is the principal one affected (Myers and Veale, 1971).

PYROGENICITY AND ANTIGENICITY

Landy: It always seemed rational to me, although specific evidence for it is lacking, that some endotoxin and/or bacteria emerge from the gut. I regard the persistence throughout life of serum antibody specificities to the entire spectrum of these enteric bacteria as evidence for this. Moreover, these antibodies are of the 19S IgM type which have a very short half life and rapid turnover. I think it not unreasonable to expect that this kind of persisting antigenic stimulus to the host is of endogenous origin and is being exerted in some pyrogenic and toxic form.

Pickering: Is the fact that the body temperature is the least variable of all the constants hitherto recorded because we all have pretty much the same flora and fauna in our gut?

Landy: No, I would be most reluctant to think that.

Cranston: If one admits that there may be a continuous "leak" of bacteria from the gut, should this necessarily have any physiological effect on temperature? If one continually infuses bacterial pyrogen, the response disappears over a period of hours (Greisman *et al.*, 1967).

Landy: The only point I am trying to make is that these bacteria or their products are being dealt with very efficiently. I mentioned the immunological data only as a form of available incontrovertible evidence of what the host must know; in contrast, there is a conspicuous absence of this in the germ-free host.

Pickering: Would there be any difference between the body temperatures of say, 20 germ-free animals and 20 animals reared in the ordinary way?

Landy: I doubt it. There were only minor differences in the physiological and immunological responses of germ-free mice and conventional mice of the same strain (Landy *et al.*, 1962) to a whole array of parameters, including injections of bacterial lipopolysaccharide, but we did not include measurement of fever. What I am suggesting is that either the host is totally walled-off from this noxious material in the gut or else this material is entering and then being dealt with very efficiently. I believe the latter is the more probable.

Pickering: Dr Palmer suggested that microbial pyrogen was a better word than bacterial pyrogen. Have moulds been shown to be pyrogenic?

Palmer: Yes. Harkness, Loving and Hodges (1950) listed about 16 moulds that produced pyrogen.

Work: The pyrogenic dose for moulds is very high; there is a tremendous dose range in the moulds from about 1 mg per rabbit to up to about 50 mg.

Snell: Dr Atkins investigated this with *Candida* (Briggs and Atkins, 1966); there are two components, as with the gram-positive bacteria. There probably is a pyrogen, extractable from the organism, that is pyrogenic in any animal, but there is also an antigen. Many animals become sensitized to moulds, as they do to bacteria, and then respond immunologically with fever to the antigen.

Pickering: So, it is an antibody-antigen reaction?

Snell: Yes, but there is also the pure pyrogen.

Pickering: Is the nature of the pure pyrogen known?

Cooper: Braude, McConnell and Douglas (1960) produced one of the mould pyrogens which differed from bacterial lipopolysaccharide in that it did not produce tolerance when repeatedly injected.

Work: Some mould pyrogens produce specific tolerance whereas many of the gram-negative bacterial pyrogens produce completely non-specific tolerance.

Pickering: Fever occurs in *Ascaris* infestation and also in trichiniasis and filariasis. Do these worms contain pyrogens? Basten, Boyer and Beeson (1970) showed that *Trichinella* infestation produces an eosinophilia but he couldn't get it by grinding up *Trichinella* and injecting it intravenously; it is a form of immune response. Therefore, I think we have to distinguish between what is pyrogenic in itself and what acts as a pyrogen by producing an antigen-antibody response.

Landy: I believe that these situations are examples of the model Dr Atkins talked about; the basic mechanism is one of cellular immunity or delayed-type hypersensitivity. The agents involved are certainly antigenic, but I doubt very much that individually they are themselves pyrogenic. However, in a sensitized subject they do evoke fever when the host is responding to the antigen in a delayed fashion.

Bangham: The continuing change of antigenic structure in cycles of malaria (Wilson *et al.*, 1969; Turner and McGregor, 1969) and trypanosomiasis (Williamson and Brown, 1964; Brown and Williamson, 1964) ties up with the immunological handling of the change of antigens released (Brown, 1969). It would be interesting to examine the isolated antigens for pyrogenic activity.

Landy: To me, what was unique and significant about Dr Atkins' report (Atkins and Bodel, 1971) was the invoking of another category of cell capable of making pyrogen. Until now the preoccupation has been with granulocytes, but we should now take into account a separate category of cells capable of secreting material that, as judged by typical host reactions, is either the same as the familiar endogenous pyrogen or is very similar.

Whittet: The viruses can also produce very high fevers.

Pickering: Is anything known about the pyrogenic substances they contain?

Bodel: In work with influenza, Atkins and Cronin (unpublished) produced fever in rabbits with both a partially-purified protein and a purified carbohydrate fraction.

Cranston: But it had to be a live virus.

Bodel: No, the fractions were not infectious.

Whittet: The fever curve from a viral infection is quite a different shape from that from bacterial pyrogen.

Landy: It is now well established that a whole series of RNA viruses which normally will not replicate on lymphocytes will do so very effectively once the lymphocyte has been activated by any of a variety of antigens; even an antigen-antibody complex can activate lymphocytes for support of viral replication.

Bangham: Has anyone demonstrated immunological tolerance to lipopolysaccharide molecules?

Landy: This was long considered impossible largely because these somatic polysaccharides are such extraordinarily potent antigens. However, it was eventually demonstrated in neonates that conventional bacterial pyrogens, given as single injections or as several, would render the neonate immunologically unresponsive as an adult.

Cranston: Were the neonates tolerant with respect to pyrogenicity as well?

Landy: Since the tests were made in mice this parameter was inappropriate. However, the mice no longer responded to the lipopolysaccharide as an antigen.

Cranston: If adult animals develop fever when given a bacterial endotoxin, having had immunological tolerance induced to the same endotoxin as neonates, this might indicate whether all febrile responses to gram-negative endotoxins are immunologically mediated.

Landy: Are you proposing that such individuals would mount a fever response to the product, but not produce antibody?

Cranston: If that were so, then it would suggest that at least some of the effects of lipopolysaccharides are not immunologically determined.

Pickering: What about the gram-positive bacteria?

Snell: This is the same situation as the model Dr Atkins worked out with *Candida*. With *Staphylococcus* there is an immunological fever mechanism in animals inadvertently immunized to the protein component, but the heat-killed cell bodies are themselves pyrogenic in all animals (Atkins, 1963*a*, *b*; Atkins and Freedman, 1963).

Bodel: Much work suggests that there are at least two mechanisms whereby gram-positive organisms can produce fever. One mechanism involves the particle itself. For example, *Staphylococcus aureus* lacking all surface antigenic components but retaining the mucopeptide residue was still fully pyrogenic; however, when that was broken down by lysozyme treatment, it was no longer pyrogenic (Atkins and Morse, 1967). The second mechanism seems to involve antigen-antibody reactions. Bacterial antigens which appear to be pyrogenic in this way include various products from staphylococci, streptococci and pneumococci, as well as tuberculin (Atkins and Snell, 1965).

Pickering: Is streptococcal endotoxin pyrogenic?

Bodel: Products released into the medium by some strains of growing streptococci are pyrogenic.

Pickering: Is what used to be regarded as streptococcal endotoxin and which was used to test patients with rheumatic fever and erythema nodosum, pyrogenic?

Snell: Kim and Watson (1965) did some work with streptococcal extracts that were pyrogenic but maybe again through an immunological mechanism.

Bodel: Schuh, Hríbalová and Atkins (1970) worked on exotoxins from haemolytic streptoccoci which are believed to be identical to erythrogenic toxin. These agents are pyrogenic apparently through a similar kind of antigen-antibody reaction.

Pickering: The most striking fever commonly seen clinically is in patients with lobar pneumonia, where onset is usually with a rigor. Within four or five hours the temperature goes up to about 104°F and during that period one can grow pneumococci from the blood. What is the mechanism of that?

Cranston: In theory the mechanism is potentially there; you have circulating pneumococci and you have phagocytosis.

Snell: The only thing missing is that you can't demonstrate the circulating leucocyte pyrogen.

Whittet: Bennet (1966) couldn't find endogenous pyrogen in the blood but he got it from the lymph.

Pickering: It is just possible that there is another mechanism of fever production besides leucocyte pyrogen.

Bangham: Perhaps more attention should be given to the monkey. If they fail to produce leucocyte pyrogen in one laboratory shouldn't this be confirmed in other laboratories?

Cranston: With monkeys, you have to make sure that this is not a specific impairment of the endotoxin response.

Myers: Some monkeys will respond to one strain of *E. coli* but not to another strain.

Bangham: For monkeys, read rabbits.

Rawlins: It would be interesting to see whether a monkey born and bred in Regent's Park developed fever when given endotoxin intravenously, because all your monkeys, Dr Myers, come from North India where they may have developed tolerance to a wide range of bacterial pyrogens.

Pickering: There are a considerable number of naturally occurring agents which will produce fever, including gram-positive and gram-negative bacteria, viruses, fungi, protozoa and worms. In some of these there seems to be fairly clear evidence that fever occurs as a result of an antigen-antibody response.

Landy: It may be more precise to refer to it as an immunologically specific reaction. It need not necessarily involve humoral or circulating antibody; it could just as well involve antigen reacting with a sensitive cell.

Pickering: But isn't that included in the antigen-antibody response? Isn't this the mechanism of the contraction of the uterus in the sensitized animal?

The antibody is supposed to be fixed on the surface of the cells and the antigen releases it.

Landy: It is certainly specific, by virtue of the cell receptor having antibody specificity, but it cannot, as a rule, be duplicated by the passive transfer of antibody as such. The capacity to react can however be transferred by sensitive cells, and this is why it is referred to as cellular immunity.

Pickering: I shall continue to talk in terms of antigen and antibody, simply because these pyrogenic agents are functioning as antigens and they act through having to have a corresponding antibody. Some pyrogens have been obtained relatively pure in the form of lipopolysaccharides and I am not sure that in them this kind of antigen-antibody response is excluded in the production of fever.

Bodel: Although the evidence is not conclusive, it seems likely that the pyrogenic activity of lipopolysaccharides normally results from some kind of antigen-antibody reaction. However, there is some evidence for an additional non-immunological "primary" pyrogenic action (Watson and Kim, 1964).

Cranston: In the rabbit with a gram-positive infection, King and Wood (1958) have shown good evidence of circulating leucocyte pyrogen.

Pickering: But it isn't so easy to demonstrate how it works on the leucocyte.

Bodel: Yes, those experiments can be done *in vitro*.

PYROGEN PRODUCTION

Pickering: We don't yet know the exact chemical nature of endogenous pyrogen and therefore we do not know whether a particular stimulus produces only one kind of endogenous pyrogen from the leucocytes of a given species. The evidence at present seems to be against the idea that endogenous pyrogen is merely the transformation of a microbial pyrogen; it seems to be something quite distinct and unrelated. As to whether it exists preformed in the cell and is simply released, I thought that some of the evidence that Dr Atkins and Dr Bodel produced was strongly in favour of actual manufacture of endogenous pyrogen, and I was very impressed with the way in which it could be cut off by inhibitors of protein synthesis, if those were put in in the early stages. Dr Bodel, have you evidence of a precursor in the leucocyte?

Bodel: There is strong evidence that a precursor appears at some early stage in the production of pyrogen, based on the effects of nucleic acid and protein inhibitors. For example, in our system, after two hours something relatively non-pyrogenic has been manufactured which is then turned into something very pyrogenic. Whether it was there all along, having been made days before in the bone marrow, or whether it was made during that two-hour interval, I don't know. However, using labelled amino acids, I have not yet been able to demonstrate synthesis of a precursor during the hours after stimulation (Bodel, 1971).

Bondy: Even that is too specific. I think we could say that some sort of protein is made during the period in which the induction occurs and after that the leucocyte pyrogen is made, but the formation of leucocyte pyrogen doesn't require protein or RNA synthesis, so presumably it represents conversion of something else to leucocyte pyrogen.

Bodel: Yes, but after two hours there is a non-pyrogenic precursor in the cell.

* * *

Bodel: I have a philosophical question about endogenous pyrogen, which is, what is it? We discovered endogenous pyrogen because we were interested in fever, but perhaps it has other important effects. Certainly cells release many different substances in response to the same stimuli as cause release of pyrogen. Could it be, for example, a kind of hormone with effects on cell types other than those in the hypothalamus, or an inflammatory agent, or an interferon?

Cooper: Philosophically we may then go on to say, why fever? Most people take the view that fever is something which aids the host; it may turn out, if one thinks teleologically, that the invader benefits more than the invaded.

Cranston: Dr Bodel's philosophical question has to wait until we know the chemical nature of leucocyte pyrogen.

Bondy: But there are certain kinetic data about this which could be examined in terms of production of other substances. For example, the question of whether interferon production would follow the same pattern of one-hour suppression and thereafter continue production would be interesting to learn.

Myers: Another philosophically important question is why is our temperature set at 37°C—why not 32°C or 42°C? Why are the other species, such as insects, not set at 37°C? Are the insects and other lower forms of life affected by pyrogen? And if not, why not?

Pickering: One thing that has become clear to everybody is that this symposium has not answered all the questions, and we can't go away and say there is nothing more to do. There is still a great deal to do, and I am sure that if we have another symposium in ten years' time the situation will look quite hopeful and quite different.

REFERENCES

Atkins, E. (1963*a*). *Yale J. Biol. Med.*, **35**, 472.
Atkins, E. (1963*b*). *Yale J. Biol. Med.*, **35**, 489.
Atkins, E., and Bodel, P. (1971). This volume, pp. 81–98.
Atkins, E., and Freedman, L. R. (1963). *Yale J. Biol. Med.*, **35**, 451.
Atkins, E., and Morse, S. I. (1967). *Yale J. Biol. Med.*, **39**, 297.
Atkins, E., and Snell, E. S. (1965). In *The Inflammatory Process*, p. 495, ed. Zweifach, B. W., Grant, L., and McCluskey, R. T. New York: Academic Press.
Basten, A., Boyer, M., and Beeson, P. B. (1970). *J. exp. Med.*, **131**, 1271–1287.
Beeson, P. B. (1938). *Proc. Soc. exp. Biol. Med.*, **38**, 160.
Bennett, I. A. (1966). *Bull. Johns Hopkins Hosp.*, **98**, 184–196.
Bodel, P. (1971). *Yale J. Biol. Med.*, in press.
Braude, A. I., McConnell, J., and Douglas, H. (1960). *J. clin. Invest.*, **39**, 1266–1276.
Briggs, R. S., and Atkins, E. (1966). *Yale J. Biol. Med.*, **38**, 431.
Brown, I. N. (1969). *Adv. Immun.*, **11**, 269.
Brown, K. N., and Williamson, J. (1964). *Expl Parasit.*, **15**, 69–80.
Greisman, S. E., Hornich, R. B., Wagner, H. N., and Woodward, T. E. (1967). *Trans. Ass. Am. Physns*, **80**, 250–258.
Hardy, J. D., Hellon, R. F., and Sutherland, K. (1964). *J. Physiol., Lond.*, **175**, 242–253.
Harkness, W. D., Loving, W. L., and Hodges, F. A. (1950). *J. Am. pharm. Ass., Sci. edit.*, **39**, 502–504.
Kim, Y. B., and Watson, D. W. (1965). *J. exp. Med.*, **121**, 751.
King, M. K., and Wood, W. B. (1958). *J. exp. Med.*, **107**, 305–318.
Landy, M., Whitby, J. L., Michael, J. G., Woods, M. W., and Newton, W. L. (1962). *Proc. Soc. exp. Biol. Med.*, **109**, 352–356.
Myers, R. D. (1971). In *Comparative Physiology of Temperature Regulation*, ed. Whitton, C. London: Academic Press. In press.
Myers, R. D., and Veale, W. L. (1970). *Science*, **170**, 95–97.
Myers, R. D., and Veale, W. L. (1971). *J. Physiol., Lond.*, in press.
Nakayama, T., and Hardy, J. D. (1969). *J. appl. Physiol.*, **27**, 848–857.
Rosendorff, C., Mooney, J. J., and Long, C. N. H. (1970). *Fedn Proc. Fedn Am. Socs exp. Biol.*, **29**, 523, abst. 1547.
Schuh, V., Hríbalová, V., and Atkins, E. (1970). *Yale J. Biol. Med.*, **43**, 184–196.
Turner, M. W., and McGregor, I. A. (1969). *Chem. exp. Immunol.*, **5**, 1–16.
Villablanca, J., and Myers, R. D. (1965). *Am. J. Physiol.*, **208**, 703.
Watson, D. W., and Kim, Y. B. (1964). In *Bacterial Endotoxins*, p. 134. ed. Landy, M., and Braun, W. New Brunswick: Rutgers University Press.
Williamson, J., and Brown, K. N. (1964). *Expl Parasit.*, **15**, 44–68.
Wilson, R. J. M., McGregor, I. A., Hall, P., Williams, K., and Bartholomew, R. (1969). *Lancet*, **2**, 201–205.
Wit, A., and Wang, S. C. (1968). *Am. J. Physiol.*, **215**, 1160–1169.

INDEX OF AUTHORS

Numbers in bold type indicate papers; other entries are contributions to the discussions.

INDEX OF AUTHORS

INDEX OF SUBJECTS

Printed in Great Britain by William Clowes & Sons Limited, London, Colchester and Beccles